Educational Guide To
U.S. SERVICE & MARITIME ACADEMIES

Educational Guide To
U.S. SERVICE & MARITIME ACADEMIES

Gene Gurney
Brian Sheehan

 VAN NOSTRAND REINHOLD COMPANY

NEW YORK   CINCINNATI   ATLANTA   DALLAS   SAN FRANCISCO
LONDON   TORONTO   MELBOURNE

Van Nostrand Reinhold Company Regional Offices:
New York   Cincinnati   Atlanta   Dallas   San Francisco

Van Nostrand Reinhold Company International Offices:
London   Toronto   Melbourne

Copyright © 1978 by Litton Educational Publishing, Inc.

Library of Congress Catalog Card Number 77-8791
ISBN: 0-442-22977-1
        0-442-22979-8 pbk.

Manufactured in the United States of America

Published by Van Nostrand Reinhold Company
450 West 33rd Street, New York, N.Y. 10001

Published simultaneously in Canada by Van Nostrand Reinhold Ltd.

15 14 13 12 11 10 9 8 7 6 5 4 3 2 1

Library of Congress Cataloging in Publication Data

Gurney, Gene.
   Educational guide to U.S. service and maritime
academies.

Includes index.
   1. Military education—United States.
2. Naval education—United States.   I. Sheehan,
Brian T., joint author.   II. Title.
U408.3.G8   355'.07'1173   77-8791
ISBN 0-442-22977-1
ISBN 0-442-22979-8 pbk.

PREFACE

With more than two million freshmen entering college each year—and the soaring costs of tuition and college living expenses—it becomes important today for every student, parent, and student counselor to carefully examine all the possibilities for obtaining a higher education at little or no expense.

For this reason, the authors have produced, for the first time in one volume, a total directory of free and partially free educational opportunities now available through United States military academies, and federally and state-financed maritime academies.

These quality educational institutions are continually on the lookout for truly outstanding students. They screen tens of thousands of applications annually, and then offer thousands of free scholarships to the most qualified.

In this important new book, students, parents, and counselors alike will find total information concerning the U.S. Military Academy at West Point; the U.S. Naval Academy at Annapolis; the U.S. Air Force Academy at Colorado Springs, Colorado; the U.S. Coast Guard Academy at New London, Connecticut; and the U.S. Merchant Marine Academy at Kings Point, New York. More than this, however, readers will discover little-known opportunities that exist at six famous state maritime academies located in Massachusetts, New York, California, Texas, Maine, and Michigan. (Representatives of each of these service and maritime academies prepared the basic information found in this book.)

There's everything a person needs to know—for example, about living accommodations, college curricula, sports, extracurricular activities, special training programs, academic requirements, coeducational opportunities, monthly government stipends, sports programs, social activities, and career opportunities for graduates of the eleven different academies.

There are copies of the application forms of each of the different academies, and practical advice on how to improve one's chances for selection to one of the coveted institutions.

In addition, photographs and illustrations show student life as it exists at the academies. Detailed maps of campus buildings and facilities make the applicant feel comfortable without ever having visited the school grounds.

But most important, this is the first comprehensive book ever published about service and maritime academies that provides a total look at all the marvelous opportunities available to college-academy-minded students. Never before has so much valuable information been condensed in such a definitive manner to aid the student, parent, guidance counselor, and others involved in finding—and obtaining—a free higher education and college degree and an opportunity to earn wages in the higher brackets after graduation.

THE AUTHORS

INTRODUCTION

Ever since my adventurous days as a West Point cadet, I have often felt there was a real need for a single great book to examine—and clearly explain—the marvelous educational opportunities available at America's outstanding service and maritime academies.

My personal concern for this grew to such a degree that, as a former Commander in the United States Strategic Air Command and Commander-in-Chief, Pacific Air Forces, I convinced two fine young military officers and authors—Lt. Col. Gene Gurney and Lt. Col. Brian Sheehan—to produce a volume that would have the honest zest, excitement of learning, and *esprit* that characterize the total educational and training experiences at our progressive service and maritime institutions of higher learning.

The dream has now reached fruition, and in this fine volume prospective military and maritime students will find a true abundance of hard factual information, practical advice, and straightforward guidance in mapping their dynamic careers.

Our nation today—perhaps more than ever before in its long history—requires dauntless leaders capable of bringing discerning intellect, reasoned wisdom, and personal courage to the solution of national defense and international maritime challenges.

The young men and women of America—upon whose future our nation's fortunes ride—will be well advised to read carefully the opportunities available to them as described in this book. They will profit immensely from the experience, and gather vast knowledge to propel them successfully toward careers in the service of the United States of America.

HUNTER HARRIS, JR.
General, USAF

CONTENTS

Preface / v

Introduction / vii

1. West Point—The United States Military Academy / 1
2. Annapolis—The United States Naval Academy / 37
3. The United States Air Force Academy / 67
4. The United States Coast Guard Academy / 104
5. The United States Merchant Marine Academy / 131
6. The State Maritime Academies / 165

The Massachusetts Maritime Academy / 170

The Maritime College of the State University of New York / 187

The California Maritime Academy / 210

The Texas Maritime Academy—Moody College of Marine Sciences and Maritime Resources—Texas A & M University / 232

The Maine Maritime Academy / 249

The Great Lakes Maritime Academy / 268

Appendix / 287

Index   299

Educational Guide To

U.S. SERVICE & MARITIME ACADEMIES

1
West Point—The United States Military Academy

Founded in 1802, the United States Military Academy (USMA) at West Point, New York, has occupied a colorful and important place in American history. Among its more than 30,000 graduates, it includes such men as Dwight D. Eisenhower, Douglas MacArthur, John J. Pershing, Ulysses S. Grant, Robert E. Lee, Frank Borman, Mike Collins, and Edwin Aldrin.

Each year the Academy graduates about 800 new officers. This is a long way from the first graduating class which numbered two men. On March 16, 1802, Congress authorized a Corps of Engineers established at West Point, New York, set its strength at five officers and ten cadets, and stated that the Corps "when so organized, . . . shall constitute a Military Academy. . . ."

The U.S. Military Academy is situated majestically on high ground adjacent to the famous Hudson River in West Point, New York.

In 1964 the authorized strength of the Corps of Cadets was increased from 2529 to 4417. The authorized number of cadets was reached by July 1972, coincident with completion of an extensive building program that provided housing and other necessary facilities for the larger Cadet Corps.

In October 1975 Congress approved, and the President signed into law, legislation directing that women be admitted to America's service academies. The admission of women to West Point in July 1976 broke 174 years of tradition, yet preserved the total integrity of the Corps of Cadets. Each cadet, male or female, is subject to the same standards of admission, training, graduation, and commissioning, except for those minimum but essential adjustments in such standards required because of physiological differences between males and females.

Two basic reasons made the founding of the Military Academy necessary. First, and obvious, was the need for a system of training officers for the army. During the Revolutionary War, the American leaders had been forced to rely upon foreign drill masters, artillerists, and trained engineers. The second reason was the ominous international political situation of 1801–02. The world had been in a turmoil for the two previous decades, and public opinion had moved quite naturally toward more energetic national government and better trained armed forces.

West Point's site has increased from the original 1800 acres purchased from Stephen Moore in 1790 to about 16,000 acres today. Fortifications, barracks, and other buildings already existed at the time of purchase because West Point had been occupied as a military post since January 20, 1778. Some remnants of the Revolutionary War fortifications at West Point remain today.

Old war fortifications remain memorialized.

Pictorial Guide to the United States Military Academy

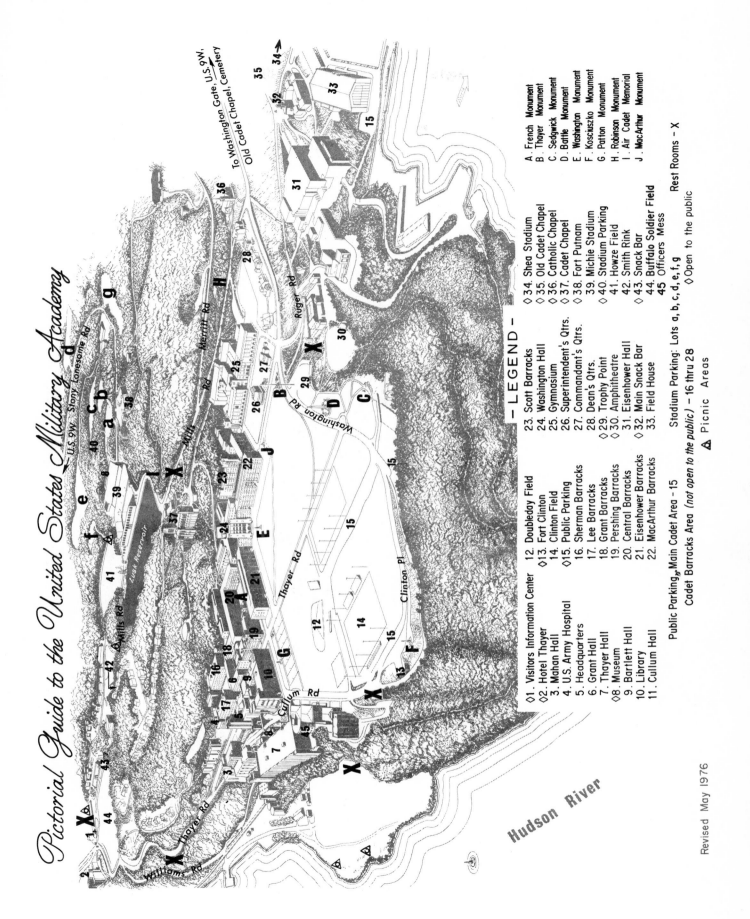

— LEGEND —

◇ 1. Visitors Information Center
◇ 2. Hotel Thayer
3. Mahan Hall
4. U.S. Army Hospital
◇ 5. Headquarters
6. Grant Hall
7. Thayer Hall
◇ 8. Museum
9. Bartlett Hall
10. Library
11. Cullum Hall
12. Doubleday Field
◇ 13. Fort Clinton
14. Clinton Field
◇ 15. Public Parking
16. Sherman Barracks
17. Lee Barracks
18. Grant Barracks
19. Pershing Barracks
20. Central Barracks
21. Eisenhower Barracks
22. MacArthur Barracks
23. Scott Barracks
24. Washington Hall
25. Gymnasium
26. Superintendent's Qtrs.
27. Commandant's Qtrs.
28. Dean's Qtrs.
◇ 29. Trophy Point
◇ 30. Amphitheatre
31. Eisenhower Hall
◇ 32. Main Snack Bar
33. Field House
◇ 34. Shea Stadium
◇ 35. Old Cadet Chapel
◇ 36. Catholic Chapel
◇ 37. Cadet Chapel
◇ 38. Fort Putnam
39. Michie Stadium
◇ 40. Stadium Parking
41. Howze Field
42. Smith Rink
◇ 43. Snack Bar
44. Buffalo Soldier Field
45. Officers Mess

A. French   Monument
B. Thayer   Monument
C. Sedgwick Monument
D. Battle   Monument
E. Washington Monument
F. Kosciuszko Monument
G. Patton   Monument
H. Robinson Monument
I. Air Cadet Memorial
J. MacArthur Monument

Rest Rooms – X

◇ Open to the public

Public Parking: Main Cadet Area – 15
Stadium Parking: Lots a, b, c, d, e, f, g
Cadet Barracks Area (not open to the public) – 16 thru 28

⚐ Picnic Areas

Revised May 1976

3

REVOLUTIONARY WAR POST

West Point was a key American fortress during the Revolution. The first known use of the term "West Point" is entered under the date of August 6, 1757, in the diary of Goldsbrow Banyar, deputy secretary of the Province of New York, who recorded, "At 7 this Evening came to an Anchor at the W. Point of Marbling's Rock (Martelaer's Rock, now Constitution Island)." In deeds, land papers, and military records of the Revolutionary period, the form used was always "the West Point," the definite article being retained because the point of reference was from the older, better known, and more populous locations on the eastern bank of the Hudson River. Usage fixed the name as West Point.

It was with the rising tide of the American Revolution that the Hudson Highlands, and West Point, intruded themselves on the consciousness of America as a region of major importance in the unfolding military and political struggle.

Left: Trophy Point at the U.S. Military Academy is a favorite vista for both cadets and visitors or to the famed site at West Point, New York.

Right: Batteries of aging cannon are permanently displayed at Trophy Point, West Point, New York.

The geological formation of the Hudson Highlands lent logic to their fortification. West Point rock is part of the belt of granite and complex Gneiss mountains stretching northeast from Pennsylvania, across northern New Jersey and southeastern New York, into western New England. In crossing the Hudson valley, these hills form a barrier of highlands about fifteen miles wide. A characteristic feature of the river's effect on the Highlands' rock is the very deep river gorge, the independent mountain masses on both sides, small rocky islands, some of which are connected with the mainland, a few terraces about 150 feet above sea level, one of which is the West Point Plain, and three remarkably sharp turns or angles in the river—at West Point, Anthony's Nose, and Dunderberg—where the hard crystalline rock has withstood the erosive power of the water. At West Point the river channel was narrowed even more by the rocky island, now Constitution Island, directly east of West Point.

One phase of the American Revolution was a contest for control of the Hudson River. In those days of water travel, the important commercial and military routes in North America were the rivers that could carry boats through and beyond the barrier of the Appalachian coastal range. The major passageway to the continent, as far as the Americans and British were concerned, was the Hudson River, which led to the Mohawk Valley, the Great Lakes, and to the interior. An immediate goal, the British felt, was to control the Hudson Valley, divide the New England colonies to the east from those to the south, and force each section to surrender separately. In order to do this, the British, who held New York City because of superior naval power, planned to sail up the Hudson River and seize several vital river crossings.

The "Battle Monument."

HIGH GROUND ON THE HUDSON

In order to wage war successfully, the Americans had to have the Hudson open for transportation of food, manpower, and munitions. West Point, the most commanding position in the Highlands, formed the sharpest angle on the river. Here the Hudson, which normally flowed in the north and south direction, turned abruptly east and then turned back again to the south. In the days when sails were used, boats were vulnerable to shore batteries when forced to slow down to navigate the turn.

Washington's army hastened to fortify the Hudson Highlands. The first group to occupy West Point was a section of a Massachusetts brigade under the command of General Samuel Holden Parsons. The unit crossed the river on the ice on January 20, 1778, and climbed the hill. The situation that faced them was not different from the more effectively publicized Valley Forge winter.

The West Point library.

5

The famous Polish general of the Revolutionary War, General Thaddeus Kosciuszko spent twenty-eight months masterminding the fortifications at West Point and is honored by this monument.

For long periods they were without pay. They were so poorly clothed that they could not work during the worst days of the winter, and their lack of an adequate and balanced diet led to disease. Many hated the inhospitable West Point rock and dubbed it "Point Purgatory." Nevertheless, the fortifications were accomplished, and now the restored works remain as visible monuments of the struggle for American freedom.

Lieutenant Colonel Louis Dashaix de la Radiere, first assigned to the task of creating the walled fortifications, found it impossible to complete his work. He was succeeded in the spring of 1778 by Thaddeus Kosciuszko who spent over twenty-eight months, almost without interruption, at West Point. Kosciuszko left his mark in the West Point defense works and in a curious monument—a little garden that he built on a terrace below the Plain to amuse himself in spare moments. The garden, with its water fountain, is still preserved.

The main fortification was at the edge of the Plain. By the summer of 1778, this work was sufficiently advanced to receive the name Fort Arnold, to honor Benedict Arnold, the hero of Quebec. Renamed Fort Clinton upon Arnold's defection, it consisted of huge trunks of trees piled up on a wall of steep rock and hand-hewn stones. Fort Clinton was supported by Colonel Henry Sherburne's redoubt, also on the level of the Plain.

Above Fort Clinton, on the high ground to the west, stood Fort Putnam whose ramparts enclosed a powder magazine, cistern, and garrison quarters. These were completed during the summer of 1779. Below Fort Putnam, covering the southern approaches, were built Forts Wyllis, Webb, and Meigs, named after their respective commanders: Colonel Samuel Wyllis, Colonel Charles Webb, and Colonel Return J. Meigs. High in the hills were four strong redoubts with connecting trails, and along the river banks were several shore batteries.

The last successful British effort to invade the Highlands took place at the end of May 1779, when they seized Verplanck's Point and Stony Point. Shortly thereafter, on July 24–25, 1779, Brigadier General Anthony Wayne, in a daring move, recaptured Stony Point with the whole enemy garrison, its cannon, and stores. Washington, who lauded Wayne's success and looked upon West Point as "the most important Post in America," then transferred his headquarters to West Point to conduct the defense. He remained there from July 25 to November 28, 1779.

General Washington's presence at West Point transformed the region into a center of attention, for next to the Congress at Philadelphia, the commanding general and his staff were the objects of continental and even international scrutiny.

The Army was practically dissolved after peace was proclaimed in 1783. A mere handful of artillerists and artificers was retained to guard such stores as were left from the Revolution: some were concentrated at Fort Pitt, but the larger portion was at West Point. The spoils of Saratoga, Stony Point, and Yorktown, including thousands of English, French, and Hessian arms were kept at West

Point and Constitution Island; and the Post was maintained as the Army's quartermaster and ordnance supply center.

The Post organization was found convenient to satisfy a number of national obligations. Captain Molly, the Margaret Corbin of Fort Washington fame, was retired to West Point so that she could get the pension and hospital stores to which she was entitled. She died near there, and her remains were later buried in the West Point cemetery.

THE EARLY MILITARY ACADEMY

Other than the short-lived attempts to organize a system of military education in the form of regimental schools during the Revolution, the earliest significant effort to establish a regular system of military instruction in the Army was expressed in the law of May 9, 1794. It was interesting in that it created the grade of "cadet" in the Corps of Artillerists and Engineers.

The title "cadet" originally meant a younger son or a younger brother. It was used in European armies to denote a young apprentice officer who was entitled to that rank because of his social status. Where primogeniture was the accepted form of inheritance, younger sons sought rank in the military services. In the United States, "cadet" was commonly accepted to mean a young gentleman with the status of a novice in the military service. In addition to authorizing the appointment of cadets, the law of 1794 also authorized the purchase of books and instruments to be used in schooling the cadets and the officers.

A monument made from a portion of the chain which was used across the Hudson to stop shipping.

Part of the Corps of Artillerists and Engineers formed the garrison at West Point, under Lieutenant Colonel Stephen Rochefontaine. Little is known about this school except that the cadets, along with the junior officers, were expected to attend regular classes in a two-story stone building called "The Old Provost," probably one of the structures on the plain built by General McDougall during the Revolution. It was later reported that some officers became indignant at descending to the grade of pupil; and, in 1796, by design or accident, the Provost, books, and instruments were destroyed by fire.

The collapse of this scheme left the War Department seriously handicapped by not having the skilled personnel to carry out its program. Responsible officers knew that knowledge acquired by experience alone could be disastrously expensive. Washington, Hamilton, and Knox, repeatedly urged that a military academy be organized.

Hamilton's proposal was embodied in the most important document concerning military policy he ever produced. In it, he described a full-blown Army school system to include a Fundamental School to train aspirants for both the Army and the Navy, as well as schools for the separate branches: a school of Engineers and Artillerists, a school of Cavalry, a school of Infantry, and a school of the Navy. Washington joined in supporting this proposal in one of the last letters he ever wrote. Congress responded with the act that Jefferson signed on March 16, 1802, establishing the United States Military Academy.

The title of the school could then have had no reference to the academic level of the curriculum to be offered; it was the accepted contemporary name given to a place of training in some special skill. "Academy" was also associated with the new schools established in America after the Revolution; it was the American academies rather than the few, tradition-bound liberal arts colleges that met the educational needs required by a new nation. Not until the third or fourth decade of the nineteenth century did "academy" become associated with an intermediate level of education. At West Point since 1802, the word "Academy," as part of the school's official title, has remained an unquestioned tradition.

The act of Congress of March 16, 1802 created a Corps of Engineers as a branch of military service separate from the Artillerists, and directed that "The said Corps when so organized, shall be stationed at West Point, the State of New York, and shall constitute a Military Academy." This law first used the phrase "military academy," perhaps as a hopeful indication of its objective, and authorized a regimental school, patterned on a medieval college, in which the Engineer officers were the masters under whom the cadets served as apprentices. This was the school in which the modern Military Academy finds its beginnings. Again, academic instruction was quite irregular, particularly because officers responsible for teaching were frequently called away for service on coastal fortifications.

Before the War of 1812, formal tuition was suspended. Up to and including this time, eighty-eight cadets had been graduated; they had entered without mental or physical examination, at all ages from twelve to thirty-four, and at various times during the year.

With the experience of previous weaknesses in mind, the Congress supplemented the law of 1802 with the legislation of April 29, 1812, and established new principles upon which the United States Military Academy has since been conducted and controlled. It authorized the appointment of professors of mathematics, natural and experimental philosophy, and engineering, who were to be in permanent residence; these were in addition to the teachers of French and drawing allowed to the Corps of Engineers by previous legislation.

The professorship in engineering was the first of its kind in any American college. By law, the overall military administration of the Post of West Point, as well as the Military Academy, was placed in the hands of an officer entitled Superintendent. A maximum of 250 cadets was authorized; and the age and mental requisites for admission were prescribed.

When the War of 1812 was over, the provisions of the law were put into effect and West Point began to change its physical appearance. By 1815, a new mess hall, an academic building, two barracks, and several brick quarters were under construction. Military practices regulated cadet life at the Military Academy.

In this period, too, the cadet gray uniform was introduced. In 1815 the cadet gray was worn in the place of the regular blue, perhaps because of the wartime shortage of dye. The change was made official by an order on September 1, 1816. Many years later, Major-General Winfield Scott said that ''cadet gray'' was adopted in honor of the victory of his gray-clad troops in the Battle of Chippewa in 1814.

In 1817, Major Sylvanus Thayer, Corps of Engineers, was ordered to take charge of the Military Academy as Superintendent.

THE ERA OF SYLVANUS THAYER

Before entering West Point as a cadet, Thayer had completed a course at Dartmouth, where he had been valedictorian of his class. Graduating from West Point in 1808, he had been an instructor at the Military Academy from 1809 to 1811, and had served with distinction during the War of 1812. In 1815 the War Department sent Thayer to France to study the military schools in that country and to collect books, maps, and instruments for West Point.

Reporting at West Point in July 1817, Thayer found most cadets absent or on special leave, regulations disregarded, studies imperfectly mastered, and the Academy in a state of disorganization. Thayer approached the problem of administering the Military Academy with two guiding principles in mind. The first principle he insisted upon making effective was strict adherence to the rule of discipline and subordination. Supporting fundamentals that had to be fostered at the Military Academy, he felt, were an ardent attachment to the military profession, a reasonable desire for preferment, and a strong and dignified regard for personal reputation.

A second principle guiding Thayer was that of advancement or promotion according to merit with no distinction between students because of financial or family background. The West Point teaching staff accepted these principles as their own, and joined Thayer in their application to the day-to-day administration of the Academy. To supervise the cadet and to conduct his military education, Thayer appointed an officer as Commandant of Cadets.

His innovations in educational methods ensured that the cadets not only learned but retained their subjects. Basically, he demanded that

West Point ceremonial uniform items include the famed Tar Bucket Saber and Sash.

United States Military Academy Coat of Arms stresses "Duty, Honor, Country."

the cadets develop habits of mental discipline and maintain standards of scholarship that have grown in importance the more they have been tested through the years.

He emphasized habits of regular study and laid down the rule that every cadet had to pass every course—any deficiency had to be made up within a specified time or the cadet would be dropped from the Corps. To carry out these rigorous standards, he limited the classroom sections to from ten to fourteen members; he rated these sections in order of merit and directed that cadets be transferred from one to the other as their averages rose or fell.

These methods and standards of Thayer's system are still used at the Academy, and Thayer's insistence on leadership integrated with excellence of character and excellence of knowledge has been the cornerstone of the Academy's training since his day. Ralph Waldo Emerson, visiting West Point in 1863, spoke of the "air of probity, of veracity, and of loyalty" the cadets had; and when, in 1898, the present coat of arms was adopted, the motto thereon, "Duty, Honor, Country," was but a later generation's attempt to put Thayer's ideal into words.

Thayer's influence extended far beyond the confines of West Point. Thayer adapted West Point to national needs by encouraging the development of the military academy so that it became the foremost American school of civil engineering. He raised the quality of the mathematics course so that, at least for several decades, West Point became the most influential mathematical school in the United States. He established West Point as the prototype of the American military school.

Thayer's most important contribution was raising the technical and ethical standard of the Army officer from that of a man with a vocational skill that could be acquired by practice to that of a man of education and training on a level with the professions of theology, law, and medicine.

For the achievements of his administration from 1817 to 1833, Thayer is celebrated by West Pointers as "The Father of the Military Academy."

THE ROLE OF WEST POINT

For their contribution to American military success, the Military Academy graduates of every generation have won many unusual tributes. The greatest praise for the early Military Academy's part in American military efficiency, as well as in American technical progress, was recorded by the historian Henry Adams, who wrote: "During the critical campaign of 1814, the West Point engineers doubled the capacity of the little American Army for resistance, and introduced a new and scientific character into American life."

General Winfield Scott, himself not a graduate of the Military Academy, testified to West Point's contribution in the Mexican War:

I give it as my fixed opinion, that but for our graduated cadets, the war between the United States and Mexico might, and probably would, have lasted some four or five years, with, in its first half, more defeats than victories falling to our share; whereas, in less than two campaigns, we conquered a great country and a peace, without the loss of a single battle or skirmish.

A similar pattern of success was reflected in the careers of Military Academy graduates on both sides of the Civil War. Until shortly before the war, no graduate of the Military Academy was a general officer in the Regular Army; by the end of the conflict, 294 graduates had become generals in the Union Army, and 151 in the Confederate Army.

For some thirty years between the Civil War and the War with Spain, although the United States was at peace, the Regular Army took part in numerous skirmishes with the Indians. In that period it is significant that twenty-one graduates won the Medal of Honor. Many of these achievements were encompassed in Theodore Roosevelt's tribute on the occasion of the Military Academy Centennial: "This institution has completed its first hundred years of life. During that century, no other institution in the land has contributed so many names as West Point has to the honor roll of the Nation's greatest citizens."

That graduates of the Military Academy served with distinction in the Spanish-American War, in World Wars I and II, in the Korean War and in Vietnam is well-established as a matter of record. Throughout its entire history, West Point has produced men prepared to assume vital roles of leadership to serve their country well and devotedly.

West Point takes pride in the immense contribution the institution and its graduates have made to national development. In the last century, Congress intended that the Military Academy should participate in furnishing the technical skills to meet the needs of a rapidly growing nation. This subsidiary mission for the Military Academy was expressed in the recommendations by Alexander Hamilton and Secretary of War James McHenry, when they advised Congress that the skills of officers educated for the military profession could also be applied to the construction of "public buildings, roads, bridges, canals, and all such works of a civil nature."

After the Military Academy was established, in effect, as the first school for the training of engineers in our country, West Point skills were applied to every aspect of internal improvement, physical expansion, and technological development. West Point men were employed on the engineering staffs of hundreds of American railroads planned or built before the Civil War from Maine to Florida and from the Atlantic coast to the Mississippi River. West of the Mississippi, they prepared surveys for the great transcontinental routes. By no means limited to railroad engineering, their work also included such widely diversified fields of engineering as river and harbor improvement, lighthouse building, road surveys, and the construction of gas, light and water systems, as well as exploration, prospecting and a

The George S. Patton Jr., monument. A modern day controversial general, who was an outstanding cadet at West Point, is honored with this statue in WWII combat dress.

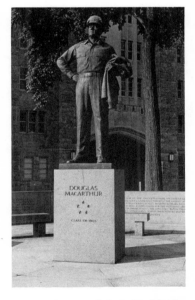

Statue of General Douglas MacArthur, a graduate of the 1903 class at West Point, is situated inside a stone memorial featuring the General's famous speech to the Cadet Corps entitled "Duty, Honor, Country."

wide variety of related activities in both public and private employment. Their investigations in the West and in Alaska brought to public attention new information on topography, geology, mineralogy, zoology, botany, and climate. Military Academy graduates were occupied in similar projects in Canada, Mexico, Cuba and Panama.

Academy cadets study sound waves in Physics II by observing the addition of two sound waves as seen on an oscilloscope.

Among the Academy graduates who won distinction in engineering work was the first graduate of 1802, Joseph G. Swift; Joseph G. Totten (Class of 1805), a leading technician in river and harbor improvements; James J. Abert (Class of 1811), headed the Topographical Engineers for more than thirty-two years; William G. McNeill (Class of 1817) and George W. Whistler (Class of 1819) achieved even greater renown in canal and railroad construction. There were few works undertaken before 1850 in connection with which the names of these men did not appear. George S. Greene (Class of 1823), who worked on the Croton Water Supply of New York City, was consulting engineer for many important municipal enterprises. The wings and the dome of the National Capitol were built under the superintendence of Montgomery C. Meigs (Class of 1836), also of Washington Adqueduct fame. George Washington Goethals (Class of 1880) was called upon by President Theodore Roosevelt to complete the Panama Canal after other engineers had failed. Leslie R. Groves (Class of November 1918) was the Commanding General of the atomic program, in 1942–1946, which not only altered warfare, but developed the means which in time will revolutionize industry and provide a better way of life for the American people and the world in general.

One may also cite with considerable detail the contribution of the

West Point faculty and graduates to the classrooms and administration of American educational institutions concerned with the mathematical sciences. By 1870, there were nineteen technical institutions in the United States of which at least ten had a direct West Point pedagogical affiliation. The successful achievement of West Point engineers influenced the growth of our nation. In no lesser degree, Military Academy graduates have won acclaim as statesmen, diplomats, churchmen, and industrialists.

THE USMA ACADEMIC PROGRAM

Today the academic curriculum consists of a core program which contains the essential elements of a broad general education, supplemented by an elective program.

The core program is designed to give the cadet an essential core of knowledge of the arts and sciences, while the elective program permits the individual to explore, in greater depth, a field in which he may have a particular interest or aptitude.

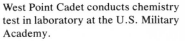

West Point Cadet conducts chemistry test in laboratory at the U.S. Military Academy.

Cadets in English class.

Elective concentrations are offered in the four areas which have substantial basis in the core curriculum. These are: the Basic Sciences; Applied Science and Engineering; the Humanities; National Security and Public Affairs.

Selection of an area of elective concentration is made in the spring of the third class (sophomore) year, and study usually begins in the second class (junior) year. Selection of one of the four areas of concentration is not required. An individual may decide to select elective courses from the entire spectrum of over 170 elective course offerings, rather than concentrating them in a specific area. Additionally, he may select courses in the interdisciplinary field of management.

Successful completion of the USMA program leads to a Bachelor of Science degree. Majors and minors are not identified as such. However, by careful selection of elective options, a cadet may complete the equivalent of a minor and, in some cases, may approach meeting the requirements for a major as defined at many institutions.

Thayer Hall serves as the Cadet academic building at the U.S. Military Academy.

Transcript credit is given for successful completion of the requirements of an area of elective concentration, but the degree is awarded without specification.

Forty-eight one-semester courses (139–142 credit hours) are required for the degree, broken down as follows:

Successful completion of the physical education program (7 credit hours);

Successful completion of the military program (6 credit hours);

Successful completion of the core curriculum. The core curriculum contains the following one-semester courses: mathematics (6 term courses); engineering fundamentals (2); earth science (1); chemistry (2); physics (3); electrical engineering (2); mechanics (2); civil, electrical, nuclear weapons systems, or general engineering (2); English (4); geography (1); foreign language (4); psychology (1); history (4); law (2); military leadership (1); economics, government, political science, and international relations (4). The core program consists of 121.5 credit hours, 57.5 or 47 percent of which are earned in the study of the social sciences and humanities, and 64 or 53 percent of which are gained through completion of mathematics, science, or engineering courses. In terms of semester courses in the core program, 21 are devoted to social sciences and humanities and 20 to mathematics, science, and engineering.

Enrollees in the basic sciences or applied sciences and engineering areas of concentration must complete the core curriculum and one additional mechanics course. They must take five of their six remaining electives in specified areas.

Learning basics of electronic circuits, three West Point cadets analyze and confirm results of experiment in the Dept. of Electrical Engineering.

Cadets who choose not to enroll in an area of concentration follow the same program as that of the basic sciences and applied sciences and engineering concentrators. There are no restrictions, however, placed on their six elective selections.

Grading is accomplished on a 3.0 scale, in which 2.0 is proficient. A grade point average is computed from term-end grades in all courses. Letter grades are assigned for transcript purposes. To obtain transcript credit for an area of concentration, a cadet's overall grade point average must be 2.250 or better.

A cadet who wishes to study a specific subject associated with an area of elective concentration may do so by pursuing one of the following elective fields associated with the areas of concentration:

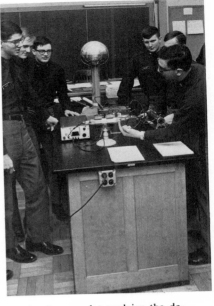

An Academy cadet explains the deflection of a beam of charged particles in a gas by an electrically charged plate during a physics lab.

Basic Science: Chemistry, computer science, mathematics, physics.

Applied Science and Engineering: Civil engineering, electrical engineering, engineering mechanics, nuclear engineering, weapon systems engineering.

Humanities: American studies, literature, Chinese, French, German, Portuguese, Russian, Spanish.

National Security and Public Affairs: Economics, geography, history, international affairs, military studies, political science.

There is one interdisciplinary field, not associated with an area of concentration—management.

Cadets practice their German in special language laboratory at West Point.

Historically 70 to 80 percent of those graduates who remain on active duty have earned graduate degrees, many times under fully-funded Army programs in an extremely wide range of disciplines:

Energy Research and Development Administration fellowships—27 (since 1964); discontinued for USMA beginning with Class of 1970;

National Science Foundation fellowships—27 (since 1962);

Rhodes scholarships—55 (since 1923);

Olmstead scholarships—33 (since 1960);

Hertz Foundation fellowships—8 (since 1973).

A faculty total of 551 men and women are assigned to academic activities at West Point, providing a student–faculty ratio of 8 to 1. Two-thirds are Military Academy graduates, and 16 percent have received their doctorates, while another 18 percent have completed their master's degrees. There are fifty-nine tenured faculty members, twenty-one professors, and thirty-eight associate professors.

Aside from Army officers, the nontenured faculty includes: 3 Allied officers, 1 Foreign Service officer, 489 Army, Navy, and Air Force officers, 6 civilians (native born linguists), a civilian visiting professor of military history, and a civilian visiting professor of English.

Professors and the Dean of the Academic Board are appointed by the President and approved by the Senate to serve until age sixty-four. Associate professors with tenure are appointed by the Secretary of the Army and serve until retirement becomes mandatory. Under current regulations such retirement occurs after thirty years of service.

Military nontenured faculty members are assigned to the Academy by the Departments of the Army, Navy, and Air Force, normally for a period of three years. Academic rank is conferred upon them by the Superintendent. Allied officers assigned to the faculty normally remain at West Point for three years; the Foreign Service officer for two years. The civilians are employed in accordance with current civil service regulations.

The Military Training Program

. . .All I want of those editors who say that "Lily-fingered cadets lounge on their velvet lawns, attend their brilliant balls and take pay for it," as I saw in a paper yesterday, is that they may go through but one plebe encampment.

This is an excerpt from a letter by then Cadet William Dutton, USMA, 1846, describing his first days at the Military Academy.

Plebe (freshman) encampment is intensive, basic, individual training designed to prepare the cadet to take his place in the Corps. This first summer is known as "Cadet Basic Training" or "Beast Barracks," and is one of the most strenuous and intensified periods of training the cadet will encounter while at the Academy.

The four years of academic and military training are designed to develop those qualities of character and physical competence needed by the officer to lead the smallest combat unit, or to advise the highest governmental council.

In the academic curriculum, standard courses provide a fundamental foundation in mathematics, science, engineering, social sciences, and the humanities. Courses are structured to promote application of this knowledge to solving problems. Advanced and elective courses offer cadets the opportunity to concentrate in areas of particular interest.

Military training provides the requisite knowledge of the professional fundamentals and doctrine, and of the basic military skills. By

Practice drilling with arms.

serving in positions of responsibility in the cadet chain of command and participating in intensive summer training, the cadet learns to apply and test those principles presented in the classroom environment.

The typical daily academic schedule (September to May) for a cadet consists of class or study from 7:50 A.M. until 3:15 P.M. on weekdays with a forty-minute break for lunch. Intramural or intercollegiate athletics, study time, parades, or extracurricular activities are held between the hours of 3:40 and 6:00 P.M. The hours of 8:00 P.M. through 11:00 P.M. are devoted to study. Taps (lights out) is at 11:00 P.M. Cadets also attend classes on Saturday mornings.

During the summer months (June through August), cadets take leave for approximately one month, and devote the remaining time to military training.

Inspection of arms.

New West Point cadets get first taste of basic military training.

During Cadet Basic Training, the new cadets receive instruction in basic military training, indoctrination, and motivation necessary to become members of the Corps of Cadets. The subjects presented are many and varied, but interspersed with intensive drill periods. There are classes, athletic competitions, marches, and all-night bivouacs during the last week of training. In the last week of August, upon return of the three upper classes to West Point, the new fourth classmen are presented to the Corps in a formal ceremony. The completion of the ceremony marks the end of the first hurdle on the path to graduation.

September marks the start of the academic year, and the fourth classmen begin their formal education, which includes mathematics, English, a foreign language, and engineering fundamentals. Classroom military subjects consist of instruction in military heritage, map reading, and basic unit tactics.

Third class summer training is devoted to enhancing soldierly skills. During this period of training, the cadet participates in "hands-on" training in such widely divergent areas as communications, weapons firing, hand-to-hand combat, and individual survival. He is exposed to the skills and techniques required by the officer

New cadets receiving instructions in the "Order of Arms."

Graduating West Point cadets march in formal parade at the U.S. Military Academy.

New women cadets at West Point are outfitted in fashionable but practical attire for student life.

serving in any of the Army's combat arms. This training is perhaps the most important tactical training in the cadet's career.

Third class academic education is expanded with the addition of physics, chemistry, psychology, and history, in addition to courses in mathematics, English, and a foreign language. Military training during the academic year builds upon the first year's instruction in basic unit tactics as employed by companies in combat.

Second class summer training is divided into two phases. The first phase consists of training in various specialized skills such as airborne, ranger, jungle warfare, and arctic warfare. The second phase consists of applying the leadership principles learned in the classroom. During this period, cadets serve as platoon leaders with combat units of the Regular Army. This phase gives cadets their first opportunity to act in bona fide leadership positions.

Academic education during this year consists of courses in electrical engineering, mechanics, physics, law, and social sciences. The

academic year's military training concentrates on combined arms operations at the maneuver battalion level.

During first class summer training one segment of the class assists in training the entering class in cadet basic training, and the other segment is assigned to perform the same function at Camp Buckner, training the new third class. This training affords each cadet an opportunity to develop the traits of leadership that will be essential to him throughout his military career.

First class academic education includes courses in engineering, psychology, English, social sciences, and history of the military art. Military training during this academic year serves as the culmination of the four years of instruction. During this year the senior cadet, using actual army units as classroom examples, participates in planning, conducting, and evaluating small unit training. He deals with situations designed to parallel those to which he will be exposed during his first five years of active service.

Upon successfully completing the four years of academic and military training, the cadet is commissioned a second lieutenant in the United States Army. Along with the gold bars, the graduate is presented a Bachelor of Science degree, ending his life as a cadet. He then begins life as an officer in the Regular Army, ready to face the many challenges and opportunities that lie ahead.

Cadets passing in review during special June Week parade at West Point.

CADET ATHLETIC AND EXTRACURRICULAR ACTIVITIES

A cadet at the United States Military Academy is one of the busiest college students in the nation. Despite a tight schedule that includes studies, classes, military training, parades, and other military functions, many cadets find time to take advantage of the numerous activities offered in the Academy's athletic and extracurricular program.

As a future officer, the cadet must be a many-faceted individual. To cope with the complexity of modern warfare and the requirements of leadership, cadets will need to acquire confidence and a working knowledge in a variety of fields. Athletic and extracurricular activities, in conjunction with the Academy's academic and military requirements, help give cadets the varied interests, knowledge, and experience in working with others needed for their future careers as leaders.

Participation in some aspect of the athletic program is an absolute requirement for each cadet. More than one-third of the Corps of Cadets takes part in intercollegiate sports during the academic year. Those who are not selected for varsity squads may join one of the athletic clubs or, not being on either a varsity or club squad, must participate in intramural athletics.

Army's best known fall varsity sport is football, but the 150-pound football, soccer, and cross-country teams invariably bring honors to the Military Academy. Winter is traditionally known as the least active season, but this is not true for most cadets. They are busy playing basketball and participating in indoor track, wrestling, swimming, gymnastics, hockey, pistol, rifle, fencing, skiing, and squash. Spring brings a full schedule of baseball, lacrosse, track, tennis, and golf.

Army cross country runner leads race against Fairleigh Dickinson trackmen in meet at West Point.

Cadet Rabble Rouser Karen Kelly
leads West Point Cadets in cheer at
Army football game.

Army basketball player shoots to
score in intercollegiate game against
Adelphi University at West Point,
New York.

The Army-Navy annual football game
is a highlight activity of the year.

Those cadets who do not compete on a varsity squad participate in intramural or club sports that help develop the physical prowess and skills that will serve them well as Army officers. Cycling, handball, judo, karate, marathon, rugby, triathlon, bowling, and volleyball provide plenty of recreation and competition. Pistol, rifle, and skeet shooting clubs develop skill with firearms. Competitive sailing and water polo teach cadets to handle boats and to take care of themselves in the water, while the Cadet Sport Parachute Club provides parachute jumping experience, and the Riding Club keeps alive the old "horse cavalry" traditions of the Army. Each of these sixteen clubs has a competitive team within its organization. Over the years, they have demonstrated their competitive ability for both team and individual participation with schools from all over the United States and with some universities in distant lands. Nearly all have gained national recognition through their participation in intercollegiate sports.

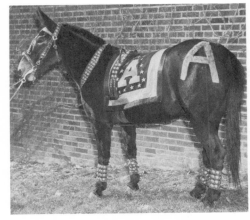

The Army mule—the mascot of the athletic teams at West Point.

Soccer has become an increasingly popular sport among cadets at West Point.

A popular sport with women cadets is horseback riding.

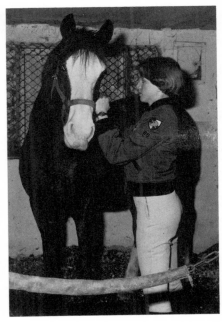

Although they do not have a varsity status, the skeet and trap teams, for example, have taken several first place wins in national competition through the sheer determination of the cadets. Cadets of the Volleyball and Judo Clubs have won individual honors for being designated All-American. Most recently, the Sport Parachute Club won their second consecutive first place in the National Collegiate Championships. In 1976 the team handball club and orienteering club won National championships for West Point. Facilities for all these activities are available to all cadets, and many participate in a different sport each season.

Physical fitness, however, is only one area in which extracurricular activities contribute to the overall program at West Point. Other clubs and organizations are designed to supplement academic study. Two of the most popular and active of these organizations are the West Point Debate Council and Forum and the Cadet Fine Arts Forum. In the former, the cadet and future officer gains practice in public speaking and the art of persuasion. He also has the opportunity to meet many of his peers from other universities throughout the United States in debate tournaments.

The many seminars of the Fine Arts Forum allow the cadet to develop his interest in the arts. Performances by nationally renowned groups and individuals offer the Corps and excellent opportunity to experience the best in the performing arts. These performances take place in the new Eisenhower Hall Auditorium, second largest on the east coast. The theater is manned by cadets who are members of the Theater Support Group, and the heavy schedule affords them the opportunity to learn the many intricacies of the day-to-day workings of the theatrical world. Also supplementing cadet studies are the Mathematics and Computer Forums, six foreign language clubs, and astronomy, electronic and engineering organizations.

Journalism is another aspect of cadet extracurricular acitivities. Cadets publish their own yearbook, *The Howitzer;* their own monthly magazine, *The Pointer;* a small handbook called *Bugle Notes,* to acquaint fourth class cadets (freshmen) with the customs, traditions, and history of the Military Academy; and *Slum and Gravy,* a sports supplement to the weekly post newspaper.

The cadet FM radio station, WKDT, is fully equipped for twenty-four-hour stereo broadcasting, and specializes in the latest music not only for the cadets, but also for other people associated with West Point. Other hobbyists can find enjoyment in the Academy's Chess and Outdoor Sportsmen's Clubs.

Student government activities include the Honor, Class, and Ring and Crest Committees. Each of the four classes is represented on these committees, which provide valuable leadership experience for cadets.

The Behavioral Science Club keeps the Corps abreast of the day-to-day problems of the world. Through "rap" sessions, guest speakers, and seminars with neighboring universities, the club discusses topics that form the substance of tomorrow's headlines. The Community Action Group of the Behavioral Science Club regularly visits patients in the local Veterans Administration Hospital and works with mentally retarded adults at Letchworth Village. On the other side of the coin is the Military Affairs Club. This club's interests range from discussions on the Volunteer Army and concepts of Army organization for the twenty-first century, to putting on a Weapons Shoot, demonstrating weapons from the time of the cave man to the present day.

For those who are musically inclined, there is the Cadet Band and Hop Bands. The nationally famed Cadet Glee Club and the Cadet Protestant, Catholic and Jewish Chapel Choirs are available for those cadets who desire to sing at official functions on the post, at religious services, or in concerts throughout the country. These singing groups often perform for national television or concert audiences.

Cadets also manage the Protestant Sunday School teaching program for post children. Another community service cadet group, The

Scoutmasters Council, sponsors scouting groups for dependents of military personnel at the Academy.

The Dialectic Society provides musical and acting outlets for cadets. Highlighting its annual activities is the time-honored "100th Night Show," presented, as the name implies, 100 nights before graduation. The show is written and produced entirely by cadets, who also provide all the actors. In addition, the Cadet Acting Troupe performs several times throughout the year.

Facilities backing this athletic and recreational program of more than 70 extracurricular activities include a fully equipped gymnasium; swimming pools, both indoor and outdoor; lakes and beaches; ice skating rink; ski slope; golf course; and all the equipment necessary to support the activities. With the completion of Eisenhower Hall, the cadet activities center, in 1974, the extracurricular activities program was further advanced by an auditorium with a seating capacity of over 4000, a large ballroom, lounges, and a restaurant that will accomodate 1000 cadets and their guests.

Approximately 100 dances, or "hops" in cadet jargon, are held during the year. Picnicking, water skiing, and canoeing in fair weather, also provide opportunities for many cadets to take a break from the strenuous requirements of their military and academic duties.

Early scenes from U.S. Army heritage are emblazoned in the Chapel window of the West Point Academy.

The West Point cadet chapel.

25

A new academic building.

West Point Cadet receives kiss and congratulations from girl friend after graduation exercises at West Point.

The three-quarter mile pathway along the Hudson River, "Flirtation Walk," widely known·from movies and television, provides cadets a temporary retreat where they and their guests may stroll and view the beauty of the Hudson River Valley.

Few cadets find life at West Point boring. Varied extracurricular activities, open to any who select them, occupy the recreational time of most cadets. During their long, hard week of academic studies, parades, military training, and athletic activities, cadets still find the opportunity to express their interests in their favorite recreation. With seventy-two activities and a yearly cadet participation of over 10,000, the extracurricular clubs and facilities are utilized extensively.

Extracurricular activities help to extend the individual cadet and to prepare him for service in the future as a Regular Army officer. Voluntary participation in the extracurricular program offers the cadet still another opportunity to develop his leadership ability and to add depth to his daily life as a cadet.

THE MILITARY ACADEMY LOOKS TO THE FUTURE

The demands made upon the Military Academy today differ from those of the last century. The current aim of the Academy is to prepare its graduates for a single profession—service in the varied tasks that the Army is called upon to perform. Deployed on three continents, the Army is in a state of transition from gunpower to atomics, from cannon to missiles, and from trucks to helicopters. New developments presage considerable changes in tactics, and in the preparation and training of an officer. At West Point, the tremendous technological advances in the arts and sciences of war are reflected in the growing emphasis upon a more broadly conceived academic and physical education program, based on a deep appreciation for the qualities of integrity and leadership. The curriculum is based on the fact that the United States Military Academy is neither a university, a liberal arts college, nor an engineering school, but a unique institution with a specific mission—that of educating and training young men and women for a lifetime career in the Army.

Cadets practice on the Parade Ground.

The lessons learned from World War II, Korea, Vietnam, and other "cold war" counterinsurgency actions have been incorporated into the tactics taught at the Academy. Training facilities were considerably expanded at nearby Camp Buckner, named in honor of General Simon Bolivar Buckner, Jr., Class of 1908. The curriculum has been broadened so that the core course requirements are now equally balanced between the arts and the sciences. Through electives, all cadets can experience a degree of specialization in the field of their choice. More and more responsibility has been given to the cadets in the administration of their organization. More privileges have been granted over the years, in evidence of the responsibility cadets have proved themselves able to shoulder. The uniforms that had been worn since Thayer's day have been joined by more utilitarian garb—shirts instead of dress coats for class, short overcoats for informal wear, and most recently, a blazer uniform consisting of gray slacks, white shirt, gold tie and black blazer.

West Point is changing, but these vast improvements have been achieved only by the striving for perfection of those who have gone before—traditionally known as "The Long Gray Line." To those graduates who, in peace and war, succeeded in molding the little fort of Revolutionary War days into the symbol and nucleus of the defense of their country—to them the Nation says, with their motto, "Duty, Honor, Country" ever before us, may we "follow close order behind you, where you have pointed the way."

June week at the U.S. Military Academy is occasion for numerous weddings of recently graduated new second lieutenants.

The Color Guard.

A traditional event, cadets throw hats in air upon graduation from the U.S. Military Academy.

Questions and Answers About West Point

Where is the Military Academy located?

The Military Academy is located at West Point, New York, fifty miles north of New York City, on the west bank of the Hudson River. The military reservation consists of approximately 16,000 acres; the West Point Post proper, framed by the Hudson Highlands, consists of 2500 acres.

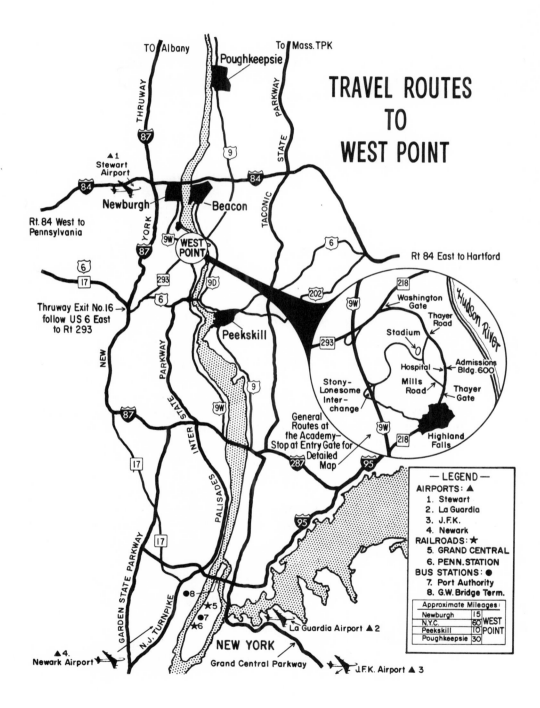

How many cadets attend the Military Academy?

The average enrollment is 4000.

What academic degree is offered?

A graduate of the Military Academy receives a Bachelor of Science degree as well as a commission as a second lieutenant in the Regular Army.

What does it cost to go to West Point?

Every cadet receives a full four-year U.S. Government scholarship, which covers tuition, room, board, and medical and dental expenses, as well as an annual salary of more than $3600, from which books, uniforms, and incidentals are purchased.

Who is eligible to attend the Academy?

Any citizen of the United States who has reached his or her seventeenth but not the twenty-second birthday, and who is not married.

How do I gain admission to the Academy?

Besides receiving a nomination, which is an authorization to be considered for admission, you must also meet specific admissions requirements.

What are the authorized sources of nomination, and how do I apply?

U.S. Senators and Representatives nominate seventy-five percent of the cadets entering the Corps. The remaining twenty-five percent are nominated by the Department of the Army to compete for service-connected cadetships. You should write to your Congressional Representative and both Senators requesting congressional nominations. For military service nomination procedures, refer to the West Point Admissions Bulletin.

When should I seek a nomination?

The best time to seek a nomination is during the spring of your junior year of high school. Some congressmen have application deadlines as early as June of the year before the entrance date of the new class. Start early.

Is it true that I must have "political pull" to be offered admission to West Point?

No. Members of Congress are interested in providing the Academy with the most outstanding young men and women. The majority of congressmen make nominations based solely on merit, as evidenced by candidate records and standardized tests.

If I receive an alternate nomination from my congressman, do I still have a chance of being admitted?

Yes. If you have an excellent record of academic and extracurricular activity, you have a good chance for admission with an alternate nomination. Each year several hundred of the best qualified alternate Congressional nominees are offered admission.

When should I seek to open an admissions file with West Point?

The best time is at the end of your junior year. Write: Admissions, USMA, West Point, New York 10996. When you have returned the questionnaire that will initiate your admissions file, West Point will begin assisting you through the nomination and qualification stages of the admissions process.

What are the admissions qualifications that I must meet?

To qualify academically, take either the American College Testing (ACT) Assessment Program or the College Entrance Examination Board (CEEB) Scholastic Aptitude Test (SAT). The ACT or the SAT must be taken by the February test date of the year of admission. Consult your guidance counselor for ACT and SAT test dates and registration procedures. Have the test scores sent to Admissions, USMA, West Point, New York 10996 and to your Congressman.

To quality medically, you must pass a qualifying medical examination, which will be scheduled by the Department of Defense Medical Review Board (DODMRB). The DODMRB is a central medical facility for all the Service Academies, and the results of one medical examination will be forwarded to each academy to which you apply.

To qualify physically, you must pass a physical aptitude examination designed to test your physical coordination, strength, and agility.

Am I required to have 20/20 eyesight to qualify for admission to West Point?

No. Your eyes need to be correctable to 20/20 with glasses or contact lenses.

How are candidates evaluated for admission?

The Military Academy uses the "whole candidate concept" of candidate evaluation to admit young men and women with balanced backgrounds in academics, athletics, and extracurricular activities. Academic potential has the most weight in this concept. Though West Point does not use cutoff scores, a look at the freshman class profile will show that most cadets score above the national average in scholastic achievement.

How difficult is the first, or "plebe," year?

This is the most physically and emotionally demanding part of the cadet's career at West Point. If you come to the Academy, you can expect a difficult first two months followed by a challenging first year. Everyone selected for admission to West Point, however, can withstand the physical and mental rigors of plebe year if he or she has the desire and determination to do so.

Do cadets ever leave the Academy?

Yes, some do. Most of those who have left, when asked if they had any advice for candidates, said that candidates should be reasonably sure they are willing to give a military career an honest try, that he or she is not coming to West Point just for prestige, because his or her parents want him or her to attend or for the degree alone.

How is the academic curriculum structured?

Four areas of elective concentration are offered within the curriculum: basic sciences, applied sciences and engineering, humanities, or national security and public affairs. A cadet may specialize in any one of these areas or can concentrate his or her electives in one of the twenty-four elective fields, such as physics, international affairs, political science, or chemistry, associated with these areas. This option will permit them to approach the degree of depth commonly associated with majors. However, a "majors" program, as commonly defined, is not available. USMA philosophy is that a broad undergraduate education will permit greater flexibility for specialization at the graduate level.

To graduate, a cadet must complete forty-eight one-semester courses, six in each of the eight terms spent at the Academy. Cadets may take up to eight electives from the more than 170 offered. In addition, there is a choice of one of six foreign language sequences, which the cadet takes during the freshman and sophomore years, and one of four engineering sequences during the senior year. Additional elective courses can be taken by validating core curriculum courses.

What is the size of the average class at the Academy?

Classes at West Point are small. A typical class section will contain about fifteen cadets. The student to faculty ratio is eight to one. Because of the small classes, instructors get to know cadets personally and can better help them cope with individual problems. Cadets can receive individual or additional instruction, on request, at any time.

What are the credentials and composition of the faculty?

The majority of the faculty at West Point are Regular Army officers who have received advanced degrees from civilian colleges and universities. Approximately ninety-eight percent have master's degrees or higher, and fifteen percent have doctorates.

How many West Pointers go on to graduate school?

Between seventy-five and eighty percent of the graduates who remain in the service attend a civilian graduate school sometime during their careers.

How much military instruction do cadets receive?

During the academic year, cadets receive two hours a week of military instruction. Most military training is conducted intensively during two of the summer months each year. During the summer following the first year, cadets receive individual training and small unit training in each of the combat arms of the Army. During the summer following the second year, cadets can opt to participate in adventure training programs, including flight training, airborne school, ranger school, northern warfare school in Alaska, or jungle school in the Canal Zone. Cadets can also spend one month as junior officers with active Army units in Germany, Alaska, Panama, Hawaii, or the continental United States. The last summer is spent visiting major Army bases (two weeks) and training new plebes (freshmen) or the yearlings (sophomores).

How much "say" do the cadets have in running things?

West Point trains leaders by requiring cadets to fill positions of responsibility and command within the Corps of Cadets. Cadets, under the general supervision of officers, manage the customs and traditions programs (with power to reward and punish); the entire cadet honor system (with power to recommend dismissal); military drill and training programs for underclassmen; the fourth class (plebe) system; the intramural athletic programs, from scheduling events to officiating at them; and all of the more than 70 extracurricular activities.

Must every cadet at West Point be an athlete?

Every cadet participates in athletics at West Point. A cadet need not be an intercollegiate athlete; however, all cadets participate in club or intramural athletics during the four years he or she is at the Academy. Any young person who passes the required medical and physical aptitude examinations will be able to compete successfully in the Academy's sports program.

What intramural sports are available?

There are three intramural seasons corresponding to the intercollegiate athletic seasons. Cadets participate in football, track, soccer, triathalon, and flickerball in the fall; basketball, boxing, handball, squash, swimming, volleyball, and wrestling in the winter; and cross-country, lacrosse, touch football, water polo, and team handball in the spring.

What other sports are offered?

Almost every sport is offered, and modern facilities and expert coaching are available; skiing, skydiving, scuba diving, and rugby are some of the cadet favorites. Cadets have distinguished themselves in twenty intercollegiate sports, including football, basketball, indoor track, swimming, wrestling, fencing, hockey, gymnastics, rifle, pistol, squash, skiing, baseball, lacrosse, track, tennis, and golf. Mountaineering, horseback riding, sailing, canoeing, flying, parachuting, judo, karate, hunting, fishing, and archery are some of the other athletic activities available.

How extensive are the athletic facilities?

The athletic facilities at West Point are perhaps the finest in the country. A large gymnasium building contains five gymnasiums, three swimming pools—one of which is Olympic size—squash, handball, and volleyball courts, and an indoor track. Varsity sports structures include Michie Stadium for football, Shea Stadium for track, Doubleday Field for baseball, Smith Rink for ice hockey, and the Field House for basketball, indoor track, and indoor rifle and pistol. Numerous athletic fields, tennis courts, and outdoor swimming facilities are located throughout the cadet area. West Point operates its own ski slope for intercollegiate, intramural, and recreational skiing as well as an eighteen-hole golf course, which is one of the finest to be found on any campus.

What are dormitory and dining facilities like?

Cadet dormitory complexes are modern, well-lighted, and comfortable. Rooms have beds, desks, closets, dressers, and sinks. There is a variety of lounges, study rooms, and recreational rooms in the dormitory areas with facilities for cadet use. All cadets dine together in the cadet dining hall. They are served high quality meals, family style.

View of cadet mess (dining room) at the U.S. Military Academy.

34

Do cadets have much time for fun and relaxation?

Cadets are busy, but they have many opportunities to relax. Competitive sports are cadet favorites, and the Army Athletic Association provides facilities for all. The cadet activities office sponsors seventy-two extracurricular activities. In addition, there are social sctivities for cadets and their dates every weekend during the academic year. A cadet activities center (Student Union), provides a 4500-seat auditorium, a large ballroom, a large snack bar, and other social and recreational facilities.

What cultural activities are offered by the Military Academy?

The Military Academy emphasizes cultural activities in an effort to produce well-rounded young officers. Noted authors, playwrights, poets, lecturers, scholars, artists, entertainers, musicians, and performing groups come frequently to address cadet audiences or perform. Mary Travers, Stonewall Jackson, Gordon Lightfoot, and Blood, Sweat and Tears, to name a few, have recently appeared at the Academy. Musical concerts by the West Point Glee Club and the West Point Band are frequent. Touring companies present live drama. A Great Films Program presents an annual series of motion picture "classics." Finally, members of the Cadet Fine Arts Forum make frequent "field trips" into nearby New York City to attend cultural affairs.

What is the dating situation?

Dating is a weekend luxury. The cadets' time during the week is at a premium, and social activities are at a minimum. However, a number of colleges are located in the West Point vicinity. Cadets often date students from these schools. A cadet hostess, whose permanent job is to manage the social and dating activities of the cadets, keeps a current file of the names of young people who have, directly or indirectly, expressed an interest in dating cadets. The cadet hostess also issues invitations, through the social directors of the local schools, to the students to attend cadet dances. More often, however, cadets invite dates "from back home" to West Point for weekend activities. The U.S. Government operates the Thayer Hotel on the West Point reservation, and inexpensive and chaperoned dormitory space is available to visiting dates; only cadets may reserve the dormitory space. Cadet dates also often stay as guests with officers' families.

Are there fraternities or sororities at West Point?

There are no fraternities, as such. However, all cadets know they belong to at least two groups, their companies and their classes. Being a part of a company of 100 cadets is similar to being a member of a fraternity or sorority. Class bonds at West Point are unusually strong and exist similarly at no other university.

An off duty moment for an evening barbecue.

Are cadets permitted to have cars?

Only first classmen (seniors) in their last semester are allowed to keep automobiles at West Point. Members of the three upper classes, and juniors, however, are allowed to drive, with the permission of the vehicle's owner, on post.

What is the Honor Code?

The Cadet Honor Code is the principal means for developing the character and integrity of the members of the Corps of Cadets and, ultimately, the officer. The essence of the honor code, that a cadet will not lie, cheat, or steal or tolerate those who do, has never changed. The code is perpetuated and administered by the cadets themselves.

How long is a graduate commited to serve in the Army?

A five-year active service obligation is incurred upon graduation from West Point.

How can a seventeen-year-old know if he or she wants to make a career of the Army?

Very few young people have their careers mapped out when seventeen years old. You should find out all you can about the opportunities of an army career and, if you are reasonably sure you want to try such a career, you should apply for West Point.

What career opportunities are available in the Army?

Upon graduation you will be commissioned in one of the four combat arms (Infantry, Field Artillery, Armor, Air Defense Artillery), one of the combat support arms (Engineers, Signal, Military Police, Military Intelligence), or, under certain circumstances, one of the many technical services. Within each branch of the Army there are many command, management, and professional career assignments and military schools. You will have the occasion to use your education and to fully develop your potential. You will be assured a variety of responsibilities; you will command troops, travel extensively, and experience challenges that your civilian contemporaries may never experience in a lifetime. Moreover, there are countless opportunities for advanced civilian schooling in a wide range of disciplines. Each officer will progress to jobs of greater responsibility based on his or her demonstrated ability.

After graduation, if I find that a military career is not for me, how worthwhile will my West Point schooling have been?

It will have been highly worthwhile. A West Point education is a broad education in the arts and sciences, designed to prepare cadets for graduate study in over one hundred fields. Because of this broad education, former cadets have done well in industrial management, engineering, medicine, law, consulting, finance, and government.

2
Annapolis—The United States Naval Academy

Through the age of sail, the age of steam, and into the age of nuclear vessels, the United States Naval Academy has had but one vital purpose: to produce career naval officers from whom will come the naval leaders and fleet admirals of tomorrow.

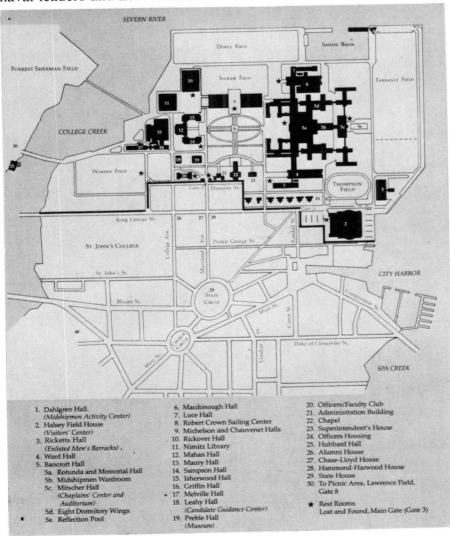

1. Dahlgren Hall.
   (Midshipmen Activity Center)
2. Halsey Field House
   (Visitors' Center)
3. Ricketts Hall
   (Enlisted Men's Barracks)
4. Ward Hall
5. Bancroft Hall
   5a. Rotunda and Memorial Hall
   5b. Midshipmen Wardroom
   5c. Mitscher Hall
       (Chaplains' Center and
       Auditorium)
   5d. Eight Dormitory Wings
   5e. Reflection Pool

6. Macdonough Hall
7. Luce Hall
8. Robert Crown Sailing Center
9. Michelson and Chauvenet Halls
10. Rickover Hall
11. Nimitz Library
12. Mahan Hall
13. Maury Hall
14. Sampson Hall
15. Isherwood Hall
16. Griffin Hall
17. Melville Hall
18. Leahy Hall
    (Candidate Guidance Center)
19. Preble Hall
    (Museum)

20. Officers/Faculty Club
21. Administration Building
22. Chapel
23. Superintendent's House
24. Officers Housing
25. Hubbard Hall
26. Alumni House
27. Chase-Lloyd House
28. Hammond-Harwood House
29. State House
30. To Picnic Area, Lawrence Field,
    Gate 8

★ Rest Rooms
  Lost and Found, Main Gate (Gate 3)

Since the Academy was founded on the banks of the Severn River in Annapolis, Maryland, in 1845, it has been fulfilling its formally stated mission: "to develop midshipmen morally, mentally, and physically to be professional officers in the naval service."

Today this three-pronged mission is accomplished in surroundings steeped in tradition, yet permeated by an atmosphere of exciting, dynamic change. The broad academic program has been tailored to meet the challenge of today's rapid pace of scientific advances and the ever-shifting tides of international affairs. The professional and military programs, too, are as sophisticated as the nuclear-powered ships and spacecraft that the midshipmen will someday command.

Where 60 young men once crowded into an old barracks of the 10-acre Fort Severn in 1845 to open the new Naval School, today, more than 4200 midshipmen, including women students, walk the 300 acres that are the Naval Academy. A brick bake house, converted to midshipmen quarters in 1856, has long since been replaced many times over by Bancroft Hall, the massive dormitory that houses the entire Brigade of Midshipmen. Modern classrooms and the latest technical equipment demand the most from today's midshipmen, and where midshipmen in 1838 set to sea in steam-powered ships, nine graduates of the Naval Academy—including Alan Shepard, Wally Schirra, James Lovell, and William Anders, Jr.—have been among the astronauts who set their course in space during the past fifteen years.

Historic view of Naval Academy and Annapolis Harbor circa 1908.

UNITED STATES NAVAL ACADEMY, 1908

Probably the most famous of the Annapolis graduates is Jimmy Carter who became President of the United States on January 20, 1977.

James Earl Carter, Jr.

PLAINS, GEORGIA

During plebe year Jimmy spent a large part of his time learning songs for the first classmen, but the only time he raised his voice after that was to shout, "Brace up!" or "Square that cap!" Studies never bothered Jimmy. In fact, the only times he opened his books were when his classmates desired help on problems. This lack of study did not, however, prevent him from standing in the upper part of his class. Jimmy's many friends will remember him for his cheerful disposition and his ability to see the humorous side of any situation

James Earl Carter, Jr., midshipman.

President Jimmy Carter

Jimmy Carter 39th President of the United States

The long transition from "sailing school" to ultramodern academic institution has been a continuing saga of challenge and achievement through thirteen decades of war and peace. However, the preeminent position now enjoyed by the U.S. Naval Academy did not simply occur: it is the outgrowth of a nation's desperate need to build a worthy naval fighting force, and man it with the best qualified officers available.

Prior to the establishment of the Naval School in 1845, the education of young officers was anything but satisfactory. Even though the system somehow produced a few remarkably famous officers, it was woefully inadequate, inconsistent, and, in many instances, farcical. Growing dissatisfaction with the old practice of training a few midshipmen on selected ships under the questionable guidance of mediocre professors of mathematics came under close scrutiny—and criticism.

AN ACADEMY IS ESTABLISHED

Unexpectedly, two developments brought about a real school for training future officers. First, there was a growing need for more professionalism in the Navy's officer corps. In the fall of 1842, a near-mutiny had occurred in the brig *Somers* while she was returning from the African coast to New York. A midshipman (who happened to be the son of the Secretary of War) and two enlisted men were hastily tried, convicted, and hanged. A public furor ensued. The resulting investigation focused unfavorable attention on the conditions in which future naval officers were being trained.

A second factor was the introduction of the steam-driven warship. The launching of the *Fulton,* the Navy's first practical ship of this kind in 1838, was followed, after only a short time, by the Ericsson-designed, screw-propelled *Princeton.* It became apparent to many that these new warships and their machinery would revolutionize the Navy—and place new and demanding requirements on its officers. A school ashore was the logical place for the Navy's midshipmen to receive the complete education to meet this challenge.

In 1845 President James K. Polk appointed George Bancroft of Massachusetts, an outstanding historian, educator, and administrator, to be his new Secretary of the Navy. With remarkable vision and skill, Bancroft almost singlehandedly established the Naval School that fall. On June 6, he wrote a note to Secretary of War Marcy requesting the use of Fort Severn at Annapolis, Maryland. On the back of his note came a brief endorsement, "I assent to the transfer, 5 August 1845, W. L. Marcy." Meanwhile, Bancroft wrote a letter to the President asking approval, which he received immediately.

Thus, the new "Naval School" came to replace four other small shore schools in New York, Boston, Norfolk, and Philadelphia, where midshipmen could voluntarily receive some education ashore. The Philadelphia shore school—the old "Naval Asylum"—was the most efficient, and supplied the new Annapolis school with much of its founding leadership.

To rehabilitate certain old buildings at Fort Severn, Secretary Bancroft placed eleven of the twenty-two professors of mathematics in the Navy on "awaiting orders" status and took their $1200 a year annual salaries for starting the school. Bancroft's unorthodox but effective administrative move enabled him to launch the infant school without going to Congress for either funds or authority.

THE ACADEMY ACQUIRES DISCIPLINE

By late summer Bancroft assembled various professors from the Naval Asylum in Philadelphia under the directorship of Commander Franklin Buchanan, a resourceful officer known as an organizer and disciplinarian. At 11 A.M. on October 10, 1845, Superintendent Buchanan called together professors and fifty midshipmen in a decrepit classroom where Tecumseh Court is now located, and declared the school open. He proclaimed a minimum of regulation, but pointed out to the young midshipmen that they were embarking upon a rare opportunity of "incalculable benefit."

At the new school, frugality and improvisation were the order of those early days. Superintendent Buchanan personally supervised the purchase of eighty beds at a small shop in the Port of Georgetown, a few miles west of Washington, D.C., for which he paid $8 apiece. He also made available $900 for the furnishing of each professor's house, "according to the taste of the occupant."

However, things did not run smoothly at the small academy; poor facilities, leaky buildings, no athletic programs, and nonexistent discipline reigned. The midshipmen were a heterogeneous group of varying ages and experience. They didn't take kindly to regimentation. They were rowdy. Some even dueled with swords, rifles, and pistols.

From one of the early dormitories, dubbed the "abbey" by midshipmen who wanted to add class to their decrepit domicile, the midshipmen tunneled out of the building, under the wall, and into town where they could purchase ale and meet the young daughters of unfriendly Annapolis families. Another dormitory of that day was "Apollo Row." It was a ramshackle structure that leaked both rain and snow. One pioneer midshipman said the rain was more objectionable "as the temperature we were able to sustain in winter with one grate fire was not sufficiently high to melt the snow."

Military discipline was not initially practiced at the Naval School. Then, in 1848, a West Point graduate and professor named H. N. Lockwood convinced the Superintendent that things were out of control. He wanted military drill initiated immediately. The school's academic board reluctantly adopted Lockwood's recommendation, and apologized in its order for such drill. The order aroused malcontented midshipmen to new heights of fury. Fortunately, their anger was only taken out in song:

A messmate before a shipmate, a shipmate before a stranger,
A stranger before a dog, but a dog before a "sojer"

Bancroft Hall as seen from the foyer of Michelson Hall.

THE CIVIL WAR CREATES PROBLEMS

These early difficulties led, in 1849, to a review of the Naval School system by a special board including outstanding officers of the day and the Superintendent of West Point. Recommendations of this board for tighter controls on the midshipmen were adopted in 1850, and the name of the Naval School was changed to "The United States Naval Academy."

One year later—in 1851—the first midshipmen practice cruise took place aboard an aging two-master on the waters of the Chesapeake Bay. A cheer arose from the midshipmen as the old sailing ship left her moorings at Annapolis, headed into the Severn River, and onto a historic voyage along the Eastern seaboard. The "middies" were back at sea. They would now be trained professionally both ashore and upon world waters.

While earlier classes had been graduated from the Naval School, it was the class of 1854 that first graduated upon the completion of a full formal course of instruction at the Naval Academy. As early as 1846, however, most midshipmen were immediately detached to sea service upon their graduation. This practice continued with the graduating class of 1854, as well as for most to follow prior to 1900.

The coming of the Civil War created a great crisis for teenaged midshipmen attending the Naval Academy from the southern states. When the states below the Mason–Dixon line began to secede in January 1861, these youngsters were forced to make decisions even mature men found difficult. If they left the Naval Academy, it meant breaking a solemn oath to defend the Constitution of the United States. On the other hand, loyalty to the government would make it necessary to renounce their families in the south and perhaps eventually, to kill some of their boyhood friends.

On the eve of hostilities, a pall hung over the Academy. There were sad farewells as the first southerners gave up their anchor and button uniforms and left Annapolis to join the Confederate States Navy. The southerners who remained did so only after painful internal debate. One of these midshipmen was Robley D. Evans of Virginia, later to be known as "Fighting Bob" Evans, one of the heroes of the Spanish-American War.

Almost immediately after the Civil War began, Annapolis was in combat territory. Because Maryland was expected to secede from the Union, the Academy gates were locked and under guard. Mobs gathered outside to hurl rocks over the wall and jeer. Midshipmen returned the missiles and insults.

THE ACADEMY GOES NORTH

Late in April the Navy decided to abandon the Academy. The *Constitution* was made ready for sea and on April 24 the battalion lined up to sing "The Star-Spangled Banner" and "Hail Columbia." Midshipmen were dismissed and given ten minutes for farewells. Then the famous old frigate was towed down the Chesapeake Bay with 200 midshipmen aboard and proceeded to Newport, Rhode Island. The *Constitution* was followed by the transport *Baltic,* which carried professors, records, books, and equipment. For the duration of the Civil War, the resort city of Newport, Rhode Island, would be the site of the Academy.

A howl was raised by critics of the Navy against the new location of the Academy. In a desperate war the vacation spot of America's wealthiest people was called a poor choice for a training site. Protestors claimed that beautiful women, extravagant parties, and other temptations would undermine the character of the young men.

Actually, Newport was a dull, uncomfortable place for midshipmen. Plebes slept in hammocks along crowded berth decks. They turned out at 6 A.M. and climbed up the rigging to the masthead and down before eating a breakfast of salt pork, hardtack, and watery coffee. Classes, study, and seamanship went on all day with breaks only for infantry drill, under a broiling sun or on a field swept by wind, snow, and sleet.

To Newport came exciting stories about recent graduates and other people well known to midshipmen. Professor Lockwood went into the Army and later commanded a brigade at Gettysburg. Commander Buchanan, first Superintendent, joined the Confederate States Navy and commanded the radical new ironclad *Merrimac,* which threatened to destroy the entire Union Navy.

While the official United States Naval Academy operated at Newport, the South formed an Academy of its own. Officers were needed to man the raiders, river boats, and harbor defenses. In 1863 the Confederate States Naval Academy opened aboard the gunboat *Patrick Henry.* The ship lay at anchor in the James River just below Richmond, Virginia. Captain William H. Parker was appointed Superintendent. He had been an honor student in the first Annapolis graduating class, and tried to pattern his new command after the U.S. Naval Academy. But the sixty Southern midshipmen, boys between fourteen and sixteen years of age, participated in more combat than studying. The Academy had a short and unsuccessful tenure.

The close of the Civil War meant the end of naval careers for all midshipmen and officers who had served the Confederate cause. Some continued to follow the sea in the American and British merchant marines. A few went abroad to work and live.

For northern officers the Civil War brought honors, promotion, and the prospect of a bright future. David Glasgow Farragut, who had led Union sea forces, was named the first full admiral in the Navy. Second in rank and destined to succeed him, was his foster brother David Dixon Porter. When the midshipmen returned to Annapolis from Newport in September 1865, Admiral Porter was named Academy Superintendent.

ADMIRAL PORTER'S "DANCING SCHOOL"

Events now moved rapidly at Annapolis. As Admiral Porter returned to the Naval Academy, he was appalled at its condition. The midshipmen's dormitories had served as hospital wards during the war. They smelled of disinfectant and death. The Superintendent's House had been a much used billiard hall and saloon, and ramshackle huts that had been soldier's beer halls spotted the parade grounds.

After cleaning up the debris, Admiral Porter set about improving the quality of life for midshipmen and faculty members alike. He incorporated some sports activity and numerous social functions into Academy life. His efforts were not without criticism, however. One newspaper, in publishing a story about the school, headlined it "Admiral Porter's Dancing School."

But the years of Admiral Porter's superintendency were the "Golden Age of the Naval Academy." Ancient subjects and outmoded regulations were discarded. The battalion was reorganized into four divisions of six gun crews each. Pay was increased to $800 a year, and a new uniform was designed. Even drill became fairly popular. A flourishing social life was encouraged.

In 1866, the renowned *Tecumseh* arrived at the Yard. This former figurement of the ship *Delaware* was originally named after the Indian Chief Tamanend. Soon midshipmen, marching with trepidation to a difficult examination, felt a bond of sympathy between themselves and the wooden image. They would salute it for good luck. Gradually a tradition was established: The wooden Indian became the "God of C," the idol for whom midshipmen gave prayers and sacrificial

The bronze statue of Tecumseh.

offerings (pennies) for success in examinations. In 1930 the original wooden statue was supplanted by a bronze replica. Today the original wooden figurehead rests on the second floor of Luce Hall, where midshipmen congregate after hours.

THE ACADEMY GROWS LARGER

Meanwhile, in 1890, athletics had become a very important part of the life at the Naval Academy. Robert Means Thompson, a graduate of the class of 1868, was influential in establishing the intramural sports programs, and intercollegiate activities. In that year the Naval Academy football team first met West Point and walloped the cadets, 24–0.

Thompson also figured prominently in other Academy affairs. As a member of the 1895 Board of Visitors, he led their recommendations condemning the inadequate facilities of Annapolis. Then, on his own, he and a New York architect named Ernest Flagg prepared a comprehensive plan for rebuilding the entire Naval Academy. This plan was approved by the Navy Department, but its implementation did not begin until 1898.

By the time of the Spanish-American War in 1898, only 207 midshipmen had graduated from the Naval Academy, yet among them was Admiral George Dewey, who led seven ships into Manila Bay and destroyed Spain's Asiatic Squadron. During the summer of the war, forty underclassmen were granted permission to join combatant ships. Among these naval cadets was a tall, slender youth from Ohio. He was Ernest J. King who later graduated fourth in his class of 1901 and went on to become a Fleet Admiral in World War II.

During the troubled years after the end of the Spanish-American War, a master plan for the Naval Academy took shape. Construction activity increased. With the French Renaissance style prevailing, Dahlgren and Macdonough Halls were completed in 1903; Isherwood Hall, the Porter Road homes for faculty members, and the main portion of the chapel were completed in 1905. Rear Admiral James H. Sands, a new Superintendent, moved into the present Superintendent's House in 1906. An academic buildings group was completed in 1907, and the wings of Bancroft Hall were finished a year later.

"ANCHORS AWEIGH"

Simultaneously, traditions were solidifying. In 1906 the twelfth playing of the Army–Navy game took place. The crowd attending that football game also heard for the first time the stirring Academy march, "Anchors Aweigh." The words for the song were written by Midshipman Alfred Hart Miles.

Just a few years later, on January 26, 1913, the Academy received the remains of the Navy's first hero, John Paul Jones, and placed them in a crypt below the chapel. The great sea commander of the Revolutionary War had died in Paris. It took a determined American

John Paul Jones.

Marble crypt of John Paul Jones is favorite point of interest to visitors to U.S. Naval Academy.

Ambassador in France years to locate his unmarked grave, for Jones had passed away in obscurity and poverty. Today the black and white marble crypt is one of the Academy's most famous and beautiful sites.

Three years later, the class of 1916 was the last to graduate on schedule before the United States went to war against Germany and Austria. The 177 men of that class were ready for what lay ahead. After the United States declared war on April 6, 1917, most of them went into troopships, cruisers, and escort craft operated by the Navy. Two million American troops were hauled to Europe by the Navy without a single loss of life. Other Annapolis men served in patrol boats, mine layers, and armed yachts. Many served in France, and some earned pilots wings for flying war missions. By the time the Armistice was signed on November 11, 1918, the men of Annapolis had again proven their courage and ability in battle.

THE CHALLENGES OF GROWTH

Continuous challenges have faced both the Academy and its graduates from the years after World War I through the present. In again modernizing its curriculum the Academy graduated its first class in 1933 with a Bachelor of Science degree along with commissions as ensigns in the U.S. Navy. During World War II the course of instruction was shortened to three years. By instituting a special schedule of summer academics, most original courses were retained. The class of 1942 was graduated in December 1941, and the classes of 1943 through 1948A (upper half of the original class of 1948) all graduated a year in advance.

Aerial view of U.S. Naval Academy at historic Annapolis, Maryland.

The Field House.

Beautiful Nimitz Library at Naval Academy illuminates Annapolis harbor at night.

For the fifth time in 100 years Academy graduates faced combat when they went forward to service during World War II. Among the heros of that great war who earlier graduated from Annapolis were Lieutenant Edward "Butch" O'Hare, class of 1937, who shot down five Japanese planes during an aerial battle in the Pacific and won the Medal of Honor; Commander David McCampbell who became the Navy's highest scoring pilot with a total of thirty-four enemy planes destroyed during a single tour of duty; Commander Howard Walter Gilmore, a great submarine combat commander; and Admirals Chester Nimitz, Thomas C. Kincaid, Daniel J. Callaghan, and Arleigh A. Burke.

While warfare in the Pacific and Atlantic was successfully waged by the U.S. Navy, significant physical expansion of the Academy grounds and buildings continued. In 1941, 22 acres were added at the foot of the slope in front of the Naval Hospital, by pumping silt from the Severn River into a steel bulkhead along the shore. A subsequent landfill operation, begun in February 1957, added a total of 53 acres of filling in Dewey Basin and extending Farragut Field to bring the yard area to 292 acres. Since the establishment of the original Naval School, 120 new acres have been added by purchase and reclamation to the original Fort Severn site.

Since then a field house was completed, in 1957, with 80,000 square feet of drill and physical training area, two additional wings were added to Bancroft Hall to provide a brigade library and assembly hall, and the Navy–Marine Memorial Stadium was completed in 1959. Other new buildings have also recently gone up on the Severn. Michelson and Chauvenet Halls, the recently completed science and mathematics buildings, are just the first of a major new construction project at Annapolis. Also newly completed are a new library, engineering building, and auditorium.

Today, traditions such as June Week, introduced under Admiral Porter, remain at the Naval Academy. White-capped midshipmen in dress blues and brass buttons still march in June Week parades. New plebes still come through the Academy gates in June and don't leave the yard again until September. Laryngitis is still common after the Army–Navy games, and drum rolls still thunder through Bancroft Hall during meal formations.

There is also new excitement of change in the continuing traditions at Annapolis. More important, however, are the broad changes in the academic and professional areas. New emphasis on broadening academic opportunities and more intense officer training have recently come about.

To accomplish these aspects of the Naval Academy's mission today, the student body is organized into a Brigade of Midshipmen. The Commandant of Midshipmen, a rear admiral or senior Navy captain, commands it. He is charged with instilling high ideals of duty, honor, and loyalty; for providing military indoctrination and physical development; and for inculcating midshipmen with the desire to achieve the high standards of performance required for midshipmen and officers of the Naval service.

Busy schedules keep midshipmen scurrying between classes at the Naval Academy.

For purposes of military training and administration, the 4200 man brigade is divided into two regiments, each of which is divided into three battalions. The six battalions are each divided into six companies. Midshipmen of all four classes are assigned to each company. Women students are assigned throughout the battalions.

THE PLEBE YEAR

When hopeful young men and women come to the Naval Academy for their initial year, they are officially known as fourth classmen, but called plebes. Plebe year is plainly tough. It makes midshipmen stand on their own feet, produce under pressure, respond promptly and intelligently to orders, and finally, to measure up to the highest standards of character, honor, and morality.

The first day of plebe summer is unforgettable. Midshipmen are given short haircuts, issued uniforms, taught marching, and served their first meal in the vast mess hall. Their military indoctrination gets off at "full speed ahead," but they are too busy to worry about it.

As summer progresses, the new middies rapidly assimilate basic skills in seamanship, navigation, and even gunnery. Infantry drill, firing an M-1 rifle under the supervision of Marine sharpshooters, sailing Navy yawls, and cruising in yard patrol craft make the youngsters proudly versatile. Team spirit and the desire to win are developed from a wide range of activities, including athletics, dress parades, seamanship, and talent shows.

Plebe summer ends in late August with Parents' Weekend, when mothers and fathers of the new midshipmen visit the Academy. A dress parade, exhibitions in sports, and the opportunity to go sailing with their sons and daughters help assure parents that their offspring are indeed taking their new lives as midshipmen in smooth stride.

Left: New plebes join together for first formation at Tecumseh Court.

Right: Dress parades are held on Wednesday afternoons at the Academy in the fall and the spring.

The cleancut new Middie.

Inspection in the ranks.

Midshipmen fall into formation in Tecumseh Court before marching to noon meal.

Soon September arrives. Upperclassmen return from at-sea training, leave, and other summer activities. The academic year cranks up. Plebe summer is officially over, but plebe indoctrination continues.

September also brings the excitement of football and other fall sports. During the football season, selected units of the Brigade travel to out-of-town games. The entire Brigade attends home games, and at

Football remains the most popular intercollegiate sport at the Academy.

Lacrosse is just one of 21 varsity and 23 intramural sports at Annapolis.

Skating in the Field House is a popular "sport" on the weekends for the middies and their dates.

. . . "Bill," a native of Ireland, is the mascot for the Navy football team. The tradition of a goat mascot was started at the time of the fourth Army-Navy game in 1893.

the end of the season, visits Philadelphia for the annual donnybrook with the Black Knights of the Hudson, the cadets of West Point.

Christmas brings a welcome two-weeks leave, which provides plebes their first opportunity to visit home since arriving at Annapolis. Classes resume in early January, followed by examinations near the end of the month. Exams are followed by a three-day leave and the start of the second semester. Five days of leave are provided over the Easter weekend.

Naval Academy midshipmen have opportunities to participate in scuba diving.

Sailing is a popular varsity and recreational sport at the Naval Academy. All midshipmen learn to sail during plebe summer.

Mixed emotions come with the approaching end of plebe year. Feelings of relief engulf most midshipmen. At the same time, confidence and pride swell in each one. A challenge has been met. Though others surely lie ahead, none may be so demanding and difficult.

Track has always been a sport in which midshipmen excel in intercollegiate competition.

The old and the new blend in the yard at Annapolis. In the background is Michelson Hall, one of the twin towers housing math and science.

THIRD CLASS YEAR

At the end of June Week (graduation week), the first year of intensive indoctrination is ended, and the new third classmen depart for two adventurous months of training on the high seas with the fleet. They are accompanied by the experienced midshipmen of the first class. This sea training is followed by a month's vacation.

With this first taste of life at sea, the midshipmen serve in many capacities, actively participating in a wide range of exciting shipboard operations. They stand deck, gunnery, operations, and engineering watches and operate ship's boats, and exercise shipboard drills.

Midshipmen get at-sea training aboard 80-foot yard patrol craft in the Severn River and Chesapeake Bay. They also spend their summers in professional training aboard ships of visiting Navy installations.

With the completion of at-sea training and summer leave, third classmen return to the Academy to start their second academic year. Although this new year brings more responsibilities in infantry drills and in watch-standing, the lessened emphasis on indoctrination leaves more time for sports and other extracurricular activities. It is a welcome and deserved change.

Following the completion of academic study during third class year and the end of their second June Week, the third classmen climb the academic ladder one more notch. The promise of sparkling gold ensign bars glitters in the foreseeable future.

SECOND CLASS YEAR

During a quick-stepping summer, this class undertakes professional studies at the Naval Academy followed by familiarization training in the four warfare specialities that comprise the naval service. At the Naval Academy the midshipman learns about submarine service through lectures and visits onboard submarines of the U.S. Atlantic Fleet. Traveling to Newport, Rhode Island, the new second classman also attends an eye-opening course at the U.S. Naval Destroyer School, and goes aboard destroyers homeported there. Flight indoctrination in naval jet training and operational aircraft at Pensacola,

Florida, later provides a knowledge of the duties of an officer choosing naval aviation as a career. Introduction to the techniques of vertical envelopment and amphibious assault, provided by the Marine Corps at one of their major training facilities, completes this summer's activity.

Following a long summer leave, second class midshipmen return to the Academy to begin their third academic year. Now greater responsibilities become theirs. Midshipmen officers are selected and trained to direct the Brigade during periodic first class absences. An important role in the indoctrination of the new fourth class is undertaken. In addition to contributing to the development of the fourth classmen, this duty makes a vital contribution to the second classman's growth as leaders. There is little time for watching the calendar. Before long, another June Week has come and gone. First class year is finally at hand.

FIRST CLASS YEAR

During this summer, all first classmen go to sea with the fleet and assume junior officer responsibilities.

On board cruise ships the first classmen stand the watches and perform the duties of commissioned naval officers. They are also exposed to the social courtesies, amenities, and customs of wardroom life. Work in navigation, watch-standing on the bridge, exercises in the combat information center, and lectures and studies on other aspects of shipboard life complete the summer's fleet training.

Well-lit, airy spaces mark the Naval Academy's new Nimitz Library.

No boys to carry the books here.

The Naval Academy Wardroom is one of the largest food operations in the world. The entire brigade can be fed in one sitting.

Midshipmen train at sea on both the east and west coasts and enroute to foreign ports. As a result, they visit a number of exotic foreign places. In recent years, ships have visited Japan, Hong Kong, and Hawaii in the Pacific; Naples, Athens, and Gibralter in the Mediterranean; and Portsmouth, Hamburg, and Oslo in northern Europe.

Meanwhile, important responsibilities are assigned to the first class in directing the Brigade of Midshipmen. Midshipmen officers, called stripers, lead the Brigade in parades, ceremonies, and daily formations. They are responsible for the conduct, military smartness, and competitive records of their units. In addition, they are in charge of the midshipmen watch organization in Bancroft Hall. The selection of three sets of midshipmen officers each academic year increases the individual opportunity for this valuable leadership experience.

In carrying out their important new tasks, the first class midshipmen find themselves calling upon all their leadership skills developed the previous three years. This final year of practical experience finds them totally prepared to assume their coming leadership role upon graduation.

A CHANGE OF COURSE

As the war in Vietnam ended, a thorough examination of the Naval Academy and its mission got underway. The main task centered around improving the quality and quantity of professional, shipboard-oriented education and training.

The Academy had to stop asking, "What must every Naval Academy graduate be able to bring to the fleet?" and had to start asking, "What must every Naval Academy class bring to the fleet?"

As part of the professional training program of the Division of Naval Command and Management, midshipmen learn the practical application of all aspects of shipboard operations. During the winter months, instruction in tactical doctrine and procedures is carried out in the Combat Information Center training rooms. Full-scale fleet tactical exercises simulated in the CIC's evoke command decisions include voice radio communications, radar presentations, air raids and tactical plots.

The answer was to produce in every graduating class a group of individual officers, all well trained in basic professional subjects, who collectively possessed a wide range of general and special knowledge and capabilities. This called for a new, thoughtful, and balanced approach to educating midshipmen.

The old core curriculum was dropped, giving each midshipman an opportunity to choose one of twenty-four majors ranging from aerospace engineering to literature to oceanography. Sixteen years ago all midshipmen took the same forty courses. Today's midshipman now has more than 400 electives available, including several courses in black studies, languages, and computer science. A minimum of 140 semester hours is required for graduation with either a Bachelor of Science or a Bachelor of Science in Engineering degree.

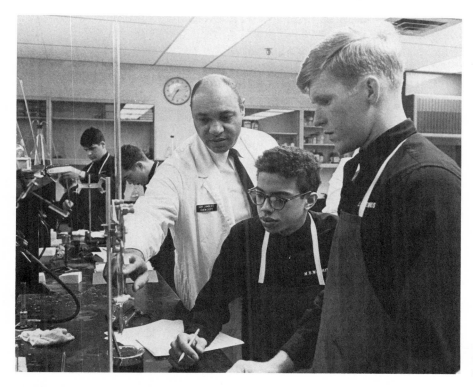

Chemistry is one of 26 majors offered to midshipmen. More than 580 courses are available.

Academy brigade marches around Tecumseh during June week parade.

For its blue-ribbon faculty, equally divided between 600 officers and civilian professors, the Academy obtained extensive laboratory and research facilities including a subcritical nuclear reactor, a subsonic wind tunnel, and an 80-foot, diesel-powered yard patrol craft outfitted with oceanographic research equipment.

More recently the Academy has been perfecting its Computer-Aided Education project (details appear in photographs), which currently involves three district programs: the IBM 1500 instructional system, teletype time-sharing, and multimedia. There are eleven courses now being taught with the aid of the computer, and the Academy may make permanent another dozen experimental computer-aided courses.

Midshipmen are required to take numerous science and engineering courses enroute to their academic degrees.

The women enjoy the classroom atmosphere.

THE TRIDENT SCHOLAR PROGRAM

A program that particularly reflects the excitement about academic expansion at Annapolis is the Trident Scholar program, which enables outstanding senior midshipmen, with personal assistance from at least one professor, to carry out independent research projects their final year. The Trident Scholar carries a reduced number of courses so that he may spend the major portion of the academic year on research and a thesis.

Studies at Annapolis have been expanded and personalized, but not relaxed. All engineering, mathematics, and science majors must still take at least five semesters of mathematics. All other majors require at least six semesters of foreign language. As one recent Academy graduate said, "There is no easy path."

The Academy also offers bright midshipmen a revolutionary validation program. In some cases, midshipmen have studied specific courses prior to entering the Academy. Rather than study a similar course again, new midshipmen can take a test covering the subject. If their scores are sufficiently high, they receive credit for the subject, and may take more advanced courses or electives.

What is the result of this new look and feel at Annapolis? Grades and interest are up, academic dropouts are down significantly, and there is a real excitement among midshipmen about the broadened academic opportunities. As one young midshipman recently said, "Flexibility is a byword in our academic program at the Academy now. It makes learning fun—and interesting."

JUNE WEEK

Inscribed in Latin above the bronze doors of the Naval Academy Chapel are the serious words, "Not Self, but Country," a motto the Academy candidates embrace from the moment they take the oath as a midshipman, through their final year at Annapolis, and onto their career in the United States Navy.

Before a first classman moves onward into professional Navy life, however, there comes a special, final week at the Naval Academy, which is the culmination of everything they have known for four years. It's a midshipman's final lesson in leadership and command. And they really get to know first hand the meaning of R.H.I.P. (Rank Has Its Privileges). The occasion: June Week!

The proud parents and sweethearts of first classmen crowd Annapolis hotels and motels for June Week events. Midshipmen in their white summer uniforms pass through the gates to mingle with the visitors. Festivity fills the air.

A leisure moment at the Naval Academy.

Parents open house and June Week are opportunities for family and girl friends to visit midshipmen at the Naval Academy.

On the first Saturday in June, graduation events get under way with a dress parade on Worden Field. The end of this day is marked by the Ring Dance, so named because the second classmen put on their class rings for the first time. By the end of June Week they will be first classmen.

The final religious service at the Academy is held the next day. Dances, concerts, the Superintendent's garden party, and scores of social activities follow in a hectic five-day period. Special honors and awards are presented to outstanding midshipmen who have excelled in everything from athletics to scholarship.

The day before graduation, parents and visitors fill the grandstands at Worden Field to witness the Presentation of Colors Parade. The national flag and brigade flags are transferred from the old to the new color guard. In the morning the Academy Band comes on the field behind the Brigade staff. The Brigade forms in lines of battalions, and the music stops. After the national anthem is played, the midshipmen captain directs the adjutant "to receive reports of the two regiments."

The receiving of reports is an old Academy tradition, which announces the year of the next graduating class. If the class of 1974 is due to graduate, the Commander of the First Regiment flashes a sword salute and calls out: "Nineteen men absent, Sir!" The Commander of the Second Regiment replies, "Seventy-four men absent, Sir!"

Midshipman kisses sweetheart prior to annual Ring Dance at Naval Academy.

The midshipman hat toss following graduation at the Naval Academy. Midshipmen receive a bachelor of science degree, and a commission as an ensign in the navy, or a second lieutenant in the marine corps when they graduate.

After the first class had made its farewell gesture to its successor, the colors are presented. Then the "Color Girl" and her midshipman escort present the flags to the incoming color guard.

With the order "Pass in Review!" the band strikes up "Anchors Aweigh" and the Brigade marches down the field past the reviewing stand. That evening the June Ball, or Graduation Ball, is held. The following day degrees and diplomas are conferred, and sad goodbyes are said as midshipmen leave the Academy for the last time to take their respected places in the active Navy.

On the Sunday after graduation day the final religious service is held in the Academy chapel.

THE BLUE AND GOLD FOREVER

An Annapolis graduate's first career opportunity comes in his or her choice of duty following graduation. The priority assigned the individual preference depends upon a number of things, including his or her class standing, the Navy's needs, and personal qualifications. Every attempt is made to assign the new officer to the duty and locality he or she requests.

Whatever the initial job, the officer will usually find that his or her responsibilities are greater than those of the civilian contemporaries. Since most Naval Academy graduates are commissioned as ensigns in the line, they are headed ultimately for command at sea. As a matter of policy, the great majority of graduates initially go to sea in a combatant surface ship—aircraft carriers, cruisers, destroyers, or on amphibious warfare ships. A certain percentage may volunteer to be selected for warfare specialities or programs such as aviation, nuclear power (either submarine or surface) and special warfare.

Prospective aviators go to sea with the fleet for six to fifteen months before entering flight training. Most of those selected for the nuclear power program go directly to nuclear power school following graduation; others follow within a few months. Graduate programs leading to advanced degrees are available to a small number of new graduates. Normally, these special programs will follow a short tour of sea duty.

A maximum of sixteen percent of Annapolis graduates may volunteer for appointment in the Marine Corps as second lieutenants, and must complete basic training before joining regular Marine units or commencing flight school.

A career at the Naval Academy is four years filled with hard academic work, demanding routine, and constant accountability, but when it's all added up, it represents one of the richest and most beneficial experiences open to the youth of America today.

Today, the Naval Academy continues to fulfill its vital mission. It will change as the times we live in change, but throughout these changes run the thread of purpose and principle that has tied together years of success in producing dedicated, courageous, and devoted officers of the United States Navy.

STEPS TO ANNAPOLIS

Nomination. Every candidate must obtain a nomination to be considered for appointment to Annapolis. The following paragraphs describe the basic qualifications all candidates must have, and provide guidance in obtaining a nomination.

Preliminary Acceptance. In July of the year preceding entry, the Bureau of Naval Personnel officially begins accepting the names of nominated candidates, and the U.S. Civil Service Commission holds the first of several competitive tests of candidates for members of congress who use this method to assist them in screening their candidates.

College Entrance Examinations. Candidates must take the College Entrance Examination Board tests (SAT and Achievement in English and Mathematics) or the American College Testing Program (ACT) tests. Tests are given every month of the year except June, August, and September. Tests must be completed before early January in order for a candidate to be considered for early notification.

Medical Examinations. Candidates must complete medical and physical aptitude examinations and insure forwarding of required documents and other information in accordance with forms and instructions sent to each nominated candidate by the Bureau of Naval Personnel.

Early Notification of Appointment. Early notification of successful candidates begins on October 15.

Nomination Deadline. January 31 is the administrative deadline for the receipt of nominations by the Bureau of Naval Personnel.

Orders to Report. On April 15, the Bureau of Naval Personnel notifies candidates of the status of their candidacy, and begins forwarding authorizations to report to the Naval Academy to successful candidates, including those who have received earlier notification of acceptence.

Admission. In early July, successful candidates report to the Naval Academy for admission as midshipmen. The date varies from year to year.

Questions and Answers About Annapolis

The following most frequently asked questions about the Naval Academy have been answered for your convenience. Greater detail is provided in the Academy's catalog, available upon request, which covers a host of subjects including admissions procedures, life at Annapolis, the curriculum, the sports program, and career options available to graduates. Careful reading of the catalog should answer almost any additional question you have about Annapolis.

Who can become a midshipman?

Admission is open to young men and women of good moral character, without regard to race, creed, or national origin. Candidates must be citizens of the United States, never married, who are at least seventeen years of age but not past their twenty-second birthday on July 1 of the year of admission.

What must I do to become a midshipman?

Obtain a nomination; qualify scholastically (acceptable test scores—either American College Testing Program (ACT) tests or College Board SAT and Achievement tests in mathematics and English; acceptable secondary school record, including college-preparatory work; top forty percent of class); meet prescribed medical and physical standards; and be selected for an appointment.

When should I apply?

Apply to your Representative and Senators for a nomination in the spring of your junior year in high school, whenever possible. If you wait until the spring of your senior year, you will not have a chance to receive an appointment to the class that enters in June of the year you graduate.

I don't know any Congressmen. How do I get a nomination?

It is not necessary to know any personally. Apply to your Congressional Representative and to both of your Senators by mail. Your application will be considered carefully. Sample letters of application are shown in the Naval Academy catalog. Each member of Congress is authorized to have five of his appointees attending the Academy at any one time. Recent legislation has authorized each Congressman to nominate up to ten candidates for each vacancy. The essential thing to remember is that, by law, you must have a nomination to be considered for appointment to the Naval Academy. Once you are nominated, you officially become a candidate and your record is then evaluated by the Naval Academy. Even if you do not receive the Congressman's appointment, if you have one of his nominations, have a good school record, and otherwise meet the basic entry standards, you will have an excellent chance to become a midshipman. In recent years, several hundred of the best qualified alternate Congressional nominees have been appointed to the Naval Academy by the Secretary of the Navy to bring the entering class up to authorized strength.

Is it difficult to enter the Academy directly from high school?

No. Nine out of ten midshipmen enter directly from high school or prep school.

My grades were about average, but I played in several sports and was student body president. Also, I had to work after school. Will these activities help me?

Yes. Evidence of leadership ability and participation in extracurricular activities, including athletics and part-time jobs, carries considerable weight in the process of evaluating candidates.

I have a high IQ and am a straight-A student. Will most of my time be spent on military subjects, or may I take any electives, such as nuclear science?

About thirty percent of the Academy's curriculum is devoted to professional military studies; but at the Academy you complete 140 units, rather than the 120 units typical of most civilian colleges. You will have a choice of twenty-four academic majors. You may specialize in nuclear science if you qualify and wish to do so. Advanced research projects are offered in many areas.

What types of majors are offered at the Naval Academy?

The twenty-four majors offered by the Academy include aerospace engineering, electrical engineering, mechanical engineering, marine engineering, naval architecture, ocean engineering, systems engineering, mathematics, chemistry, oceanography, applied science, physics, analytical management, general management, operations analysis, European studies (French, German, or Italian), Far Eastern studies (Chinese), Latin American studies (Spanish or Portuguese), Soviet studies (Russian), economics, American political and military systems, international security affairs, literature, and history.

Do I get to choose any of my courses?

Yes, you will choose your major and the majority of your courses.

How much does it cost to be a midshipman at the Academy?

Tuition, room and board, and medical and dental care are provided by the government. Midshipmen receive a $225.30 monthly salary for uniforms, books, and personal needs. A $300.00 deposit is required on entry.

What is my military obligation on graduation?

The total military service obligation of an Academy graduate is six years. Current directives require five of these to be on active duty as a commissioned officer in the Navy or Marine Corps.

I have a nomination to both the Naval Academy and the Air Force Academy. Is it necessary that I undergo two medical examinations?

No. A medical examination conducted by any one of the services for its service academy is acceptable to the others. However, you must ensure that the examining center forwards the examination results to the appropriate Academies.

I am in college now. Is it too late to enter the Academy?

No, as long as you will not have passed your twenty-second birthday on July 1 of the year of admisssion. Prior college work will permit study of advanced courses at the Academy. Normally, about ten percent of the members of each entering class have been in a civilian college prior to entering the Naval Academy.

My father was in the Armed Forces. Will this help me to get a nomination?

Children of career members of the regular and reserve forces who are on active duty or who are retired may be considered for a nomination under the Presidential category.

If I am eligible for both Congressional and Presidential nominations, can I request nominations from more than one source?

Yes, and you should. The more nominations you have, the better chance you have to enter the Academy.

What part of the medical examination gives the most difficulty to candidates?

The eye examination. Visual acuity of 20/20 is required. However, a limited number of outstanding candidates may be granted waivers for visual acuity that is no worse than 20/100, correctable to 20/20.

Where do midshipmen live?

They are housed in one building, Bancroft Hall, the largest dormitory in the world. Bancroft Hall has more than 4.8 miles of corridors and 33 acres of floor space. Each room has its own shower and all rooms have been remodeled in recent years. The mess hall is also located in Bancroft Hall. Here all 4200 midshipmen are able to sit down and eat at one time. The food is served family style. Bancroft Hall is named after George Bancroft, who was Secretary of the Navy when the Naval Academy was founded. The Hall contains a store, medical and dental facilities, a soda fountain, bowling alleys, and numerous other recreational facilities.

How much social life would I have at the Academy.

Social life is limited during the first year. After the initial (plebe) year, there is a wide range of social activities available. There are weekend dances and other extracurricular activities at the Academy, and there are opportunities for afternoon liberties in town and for a number of weekends and longer leave periods away from the Academy.

Do I get to fly?

All midshipmen receive flight instruction in naval aircraft during their third summer. Selected volunteers receive flight training after graduation, leading to designation as Navy or Marine aviators.

I don't like sports. Do I have to try out for anything?

Yes. If you really dislike sports, the Academy's varsity and intramural athletic programs may not appeal to you. Every midshipman is required to participate in athletics, either varsity or intramural, for the development of his character, physical fitness, and competitive spirit.

What sports are available at Annapolis?

Just about any one you could want (many are offered at both varsity and intramural levels) including football, boxing, handball, cross-country, soccer, rugby, sailing, basketball, fencing, gymnastics, pistol, rifle, squash, volleyball, swimming, water polo, track, wrestling, baseball, softball, crew, golf, lacrosse, and tennis.

How about extracurricular activities?

There are more than sixty-five choices. A partial listing includes the Glee Club, the Antiphonal Choir, the Protestant and Chapel Choirs, the Drum and Bugle Corps, the Concert Band, five separate combos, and a dance band, (the NA-10), staffs of the yearbook (*The Lucky Bag*), two magazines (the *Trident* and the *Log*) and *Reef Points,* an introductory booklet for freshmen; theater groups (the Masqueraders and the Musical Clubs); such organizations as the Photographic Club, Gun Club, Amateur Radio Club, Scuba Club, Art and Printing Club, Brigade Activities Committee, Public Relations Committee, and the midshipmen's radio station WRNV, a chapter of the American Institute of Aeronautics and Astronautics, the physics honor society, the management society, and the oceanography society; and, finally, there are sailing and the activities of the Yard Patrol Squadron, a group of midshipmen interested in perfecting their professional skills afloat aboard the Academy's six motorized yard patrol craft.

3
The United States Air Force Academy

The basic stimulus for the United States Air Force Academy was the evolution of American military aviation. When World War I clearly demonstrated the growing importance of the air arm, there were several proposals for establishing an American aviation academy. Nothing concrete developed, however, although some responsible national leaders recognized the need for an educational program to prepare men for the profession of aerial warfare, as the United States Military Academy and the United States Naval Academy prepared them for ground and sea warfare. The vital role of air power in World War II again stimulated several proposals in Congress for the establishment of an Air Academy. However, all of them were dropped pending the conclusion of the war.

In October 1946, the U.S. Army Air Force itself prepared legislation to establish an Air Academy along the lines of West Point and

Located at the foothills of the Rocky Mountain Rampart Range just north of Colorado Springs, sits the U.S. Air Force Academy. Instruction is given in academics, athletics, airmanship and military studies to more than 4,000 cadets who, upon graduation, will receive bachelor of science degrees and commissions as air force second lieutenants.

Annapolis. The Academy legislation was held in abeyance, however, until final passage of the armed services unification proposals. In the meantime, the Air Force made agreements with the Army and Navy to commission a certain percentage of West Point and Annapolis graduates each year as Regular Air Force officers; the agreements were to hold until the Air Force could secure its own academy.

The post-World War II movement to establish an Air Force Academy was given impetus by the strong belief of Air Force officers and many civilians in the vital role exercised by air power, not only in warfare, but in the maintenance of continuing national security.

There were other important reasons for the establishment of the Academy, too. It would help inculcate the concept of service, which should govern the lives and actions of future air commanders. It would help satisfy the need for a solid core of regular junior officers well grounded in military, scientific, and social studies, in the humanities, and in the fundamentals of military training. No service academy or civilian institution offered a balanced curriculum in these subjects that could serve as a foundation upon which to build future Air Force careers. Also, none of them furnished the proper motivation for a young man to desire a lifetime career in the Air Force.

AN ACADEMY IS ESTABLISHED

Interest in an Air Force Academy soon began to manifest itself in several ways. Senators and Representatives introduced a large number of bills calling for the establishment of an academy at specific sites. Other individuals and groups interested in air power called for the creation of an academy. On September 1, 1948, General Hoyt Vandenberg, Chief of Staff of the Air Force, directed plans for an air academy based upon a four-year course of instruction, generally along the lines of the existing service academies. An Air Force Academy Planning Board was established. It centered its operations on the answer to the question: what does an Air Force officer need to know in an age of supersonic jets, long-range guided missiles, and nuclear weapons; in an era when man stands at the threshold of space; in a period of rapid social change and of great political decisions?

It was agreed that the Air Force Academy should not be just an institution devoted to training junior officers, but should provide the educational background required by senior air officers of a nation that had assumed leadership of the noncommunist world.

In March 1949, while the Air Force Academy Planning Board was carrying on its studies, Secretary of Defense James Forrestal appointed the Service Academy Board to survey the whole question of educating career officers for the armed forces. In its report, the Stearns-Eisenhower Board recommended that West Point and Annapolis should continue, as they had for more than a century, to furnish career officers for the Army and Navy. The Board further recommended that a similar institution be established to furnish the Air Force with a continuous flow of qualified young career officers.

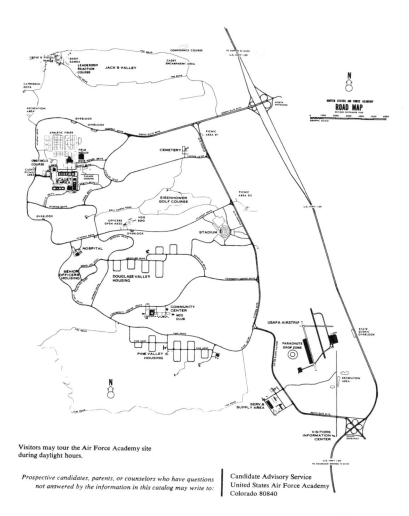

Visitors may tour the Air Force Academy site
during daylight hours.

*Prospective candidates, parents, or counselors who have questions
not answered by the information in this catalog may write to:*

Candidate Advisory Service
United States Air Force Academy
Colorado 80840

The 17 spires of the Academy inter-faith chapel dominate the center of this aerial photograph of the Academic complex of the U.S. Air Force Academy. The long building in the background is Harmon Hall, the administration building. The taller windowless one at the extreme right is Arnold Hall, the cadet social center.

In December 1949, Lieutenant General Hubert R. Harmon was named Special Assistant to the Chief of Staff for Air Force Academy Matters. General Harmon was charged with all planning for the future Air Force Academy, including site, facilities, equipment, building, curriculum, etc. Not the least of his duties was to obtain the proper legislation from Congress.

General Harmon prepared a program of instruction for the proposed Academy. He was assisted by officers who were specialists in various curriculum subjects. They utilized the Academy Planning Board recommendations as a basis for a proposed curriculum. These recommendations were studied, analyzed, revised, and refined, both by the planning staff and by civilian consultants from leading universities. The scientific courses were reviewed by faculty members of the Massachusetts Institute of Technology, and the social sciences and humanities courses by the faculties of Stanford and Columbia Universities.

The outbreak of the Korean War delayed plans for the Air Force Academy. In January 1954, however, Congress began hearings on HR 5337, a Defense Department-sponsored measure to establish such an academy. Both branches of the Congress completed final and favorable action on March 29, and the President approved the Act as Public Law 325, 83rd Congress, 2d Session, on April 1, 1954.

On July 27, 1954, under the authority of Public Law 325, the United States Air Force Academy was established as a separate agency under the Chief of Staff of the Air Force. General Harmon assumed command of the Academy as Superintendent, on August 14 of the same year. He and his growing staff then busied themselves with the enormous tasks involved in putting the Academy into operation: assembly and organization of the faculty; final preparation of the educational program; development of the physical facilities; and all the procedures connected with the selection and admission of the first class of cadets.

The Academy's dedication ceremony took place at Lowry Air Force Base in Denver, Colorado—temporary home of the academy—on July 11, 1955. Present were the Secretaries of the Army and the Air Force, and distinguished military, civic, and educational leaders. In the dedicating address, the Secretary of the Air Force, Harold E. Talbott, stated:

Flying, especially military flying, makes great demands on character. A man aloft in the great wilderness of space knows a loneliness of the spirit as well as of the body. There is no hand to touch, no shoulder on which to lean for even a moment's respite. There is no one from whom to seek advice, no one to share responsibility. Only the compulsion and discipline of duty drives a man to the completion of his task. So it is of the man, and not the machine, we must think of when we speak of air power. Thus, it is to the human element that the Air Force Academy is dedicated, and especially to the leadership we must have if our country is to survive. . . .

The first academic class, 306 new cadets, had reported on the

MAN'S FLIGHT THROUGH LIFE IS SUSTAINED BY THE POWER OF HIS KNOWLEDGE.

Cadets at the Air Force Academy present the colors during a full dress parade.

En route to afternoon classes, a U.S. Air Force Academy cadet runs past the Eagle statue. During the first half of their freshmen year, cadets run to all classes and formations.

Air Force Academy cadets in formation during a review ceremony.

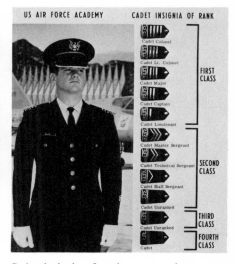

morning of Dedication Day. They took part in the ceremonies and then began their training.

Since then the mission and objectives of the United States Air Force Academy have remained the same. The Academy is to provide

Cadet insignia of rank as worn by cadets at U.S. Air Force Academy.

"instruction and experience to each cadet so that he graduates with the knowledge and character essential to leadership and the motivation to become a career officer in the United States Air Force." Inherent in the mission are the objectives of the Air Force Academy. As a minimum requirement, the objectives are to provide each cadet with basic baccalaureate-level education in airmanship, related sciences, the humanities, and other broadening disciplines; a knowledge of, and an appreciation for, air power, its capabilities and limitations, and the role it plays in the defense of the nation; high ideals of individual integrity, patriotism, loyalty, and honor; and a sense of responsibility and dedication to selfless and honorable service.

In order to fulfill its mission and objectives, the Air Force Academy offers a four-year program of instruction. The objectives of the service academy system, of which the Air Force Academy is a part, transcend nearly all the normal patterns found in other American institutions of higher education. They also introduce military goals that are not found in many other schools. The curriculum, therefore, consists of a diversified range of courses designed to prepare the cadet for a broad scope of activity as an Air Force officer. It is divided into three phases of cadet education: an academic program supervised by the Dean of the Faculty, a leadership and military training program under the direction of the Commandant of Cadets, and an athletic and physical education program directed by the Director of Athletics. The successful completion of the curriculum entitles the cadet to graduate with an accredited bachelor of science degree and a commission as a second lieutenant.

THE ACADEMY PROGRAM

About eighty percent of a cadet's academic effort is devoted to a balanced core curriculum, which provides a general and professional foundation appropriate for any citizen and essential to a career Air Force officer. Part of this work will be found in the general education programs of most undergraduate institutions, as prerequisites and foundations for further specialization. Beyond these fundamental requirements, however, the special needs of future Air Force officers are met by professionally-oriented courses, such as human physiology, computer science, economics, military history, and astronautics. The core curriculum includes nineteen and one-half courses (58½ semester hours) in mathematics, science, and engineering, and seventeen and one-half courses (52½ semester hours) in social sciences and humanities.

Building on the foundation of the core curriculum, each cadet has the opportunity to select about from twenty-seven to thirty-three semester hours of elective courses for one of the twenty-three subject-area majors offered by the Academy. These majors have been designed with the interests of the cadets and the needs of the Air Force in mind, and each cadet is required to complete a major as part of his course of instruction. About fifty percent of the cadets

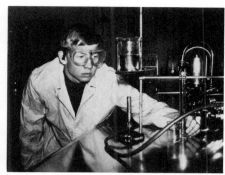

Chemistry student in laboratory at Air Force Academy conducts experiment.

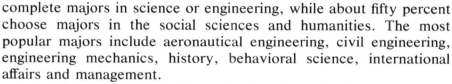

Cadets engaged in classroom work. With a core curriculum heavily weighted to science and engineering, a cadet may select a major from one of 22 academic areas.

complete majors in science or engineering, while about fifty percent choose majors in the social sciences and humanities. The most popular majors include aeronautical engineering, civil engineering, engineering mechanics, history, behavioral science, international affairs and management.

A key feature of the Air Force Academy's academic program, and an innovation in service academy education, is the enrichment program, formally established in 1956. The program has one basic objective: to challenge each cadet to advance academically as far and as fast as he can. It takes into account the differences in ability, preparation, and interest of the cadets, and offers each cadet the challenge to live up to his potential in the pursuit of academic excellence.

Each cadet may participate in this voluntary program in several ways: by transfer of credit for college courses taken prior to enrollment at the Academy; by successful completion of validation examination at the Academy; by enrollment in advanced placement or accelerated courses, especially in the basic sequences in mathematics, chemistry, or physics and finally, for cadets who maintain a satisfactory grade average, by completion of "overload" courses.

The Academy's 550-man academic faculty consists mostly of military officers. Members of the faculty must play a dual role in the classroom; in addition to imparting knowledge, they must assist with the development of character, motivation, and qualities of leadership essential to Air Force career officers. Military members of the faculty

Electronics—along with other science courses—is a major focal area of study for cadets at the U.S. Air Force Academy.

Cadets have opportunity to work with computers in science technology program at Air Force Academy.

All cadets are required to take 10 hours of language as a part of humanities studies. This is a class in Russian.

A class in Physiology studies reflex actions of the heart.

are selected from career officers, all volunteers, who have established outstanding records of performance and career dedication.

In most cases they have military professional experience related to their academic specialities. In all cases they hold at least a master's degree; about one-third hold a second professional degree, and more than one-fourth have earned doctorates.

WOMEN IN THE AIR FORCE

The contributions made by women in the armed forces are not new. Women have long served in the Nurse Corps of the various services. During World War II, women served in the air forces as part of the Women's Army Corps (WAC), the Army Air Forces, and Women Air Service Pilots (WASPS).

These contributions were recognized by Congress when it passed the Women's Armed Forces Integration Act in 1949. This act recognized women as a prominent part of the armed forces, and created the Women in the Air Force (WAF) as a segment of the United States Air Force.

Under today's equal personnel concepts, women are not organized as a separate corps, or referred to as WAFs, but form an integral part of the Air Force. They are trained and assigned under essentially the same policies as men, and they compete equally with men for promotions.

The Officer Training School and the Air Force Reserve Officer Training School and the Air Force Reserve Training Corps (AF-ROTC) have been open to women for several years. One of the final achievements of integrating women into Air Force training programs was made possible on October 7, 1975, when President Ford signed into law a bill authorizing admission of women to the national service academies. The law states that the standards required for admission, training, graduation, and commissioning of women will be the same as those required for men, except for minimum adjustments in standards required because of physiological differences between men and women.

One of the 155 women who entered the Air Force Academy in 1976 prepares to leave her luggage and sign in for four years.

You have to pay close attention in the German class.

The cadets also have to learn the basics of drilling.

The obstacle course.

RESEARCH AT USAFA

Meanwhile, cadets, faculty and an Air Force tenant organization have joined forces at the Air Force Academy to develop a research program with tangible, immediate benefits to the Air Force careers of those participating. Although research always has been an important part of the academic life of any university, it has received increased emphasis in recent years, because rapidly advancing technology makes it necessary for faculty members to pursue a continuing program of self-education. Research contributes to the faculty's self-development, and the results of research may be transmitted directly in the classroom. Ultimately, the country's scientific ability is improved through such research programs. In some cases, faculty members have been able to complete a doctoral dissertation only by continuing research after they have reported for duty.

The objective of the Academy's research program for cadets is to expose them to an environment wherein they have an opportunity to be creative, to develop their intellectual curiosity, to search out and solve problems, and to relate the various disciplines that comprise a liberal and technical education. It also provides selected cadets the opportunity to participate in an advanced research environment, and to begin a graduate program.

An important part in the Academy's research program is played by the Frank J. Seiler Research Laboratory of the Air Force Office of Aerospace Research. This laboratory is an in-house research facility in the physical sciences, primarily in general chemistry, physical chemistry, fluid dynamics, flight mechanics, and applied mathematics. Because of the Seiler Laboratory's location here, the Academy is able to draw upon its research talents and equipment to supplement the research efforts of the cadets and members of the faculty.

First class cadets may spend part of their summer working with the Air Force Systems Command, Office of Aerospace Research, and National Aeronautics and Space Administration. The cadets work on such research projects as incentive contracting, the effects of plane waves in solids, shielding against magnetic fields, the production of continuous emission by lasers, the evaluation of an integral in scattering theory, a theoretical analysis of electron emission, the solution of simultaneous equations arising from acoustic wave scattering, and the determination of Brillouin zones in a sulfide crystal, to name a few.

The summer work allows the cadets to gain an understanding of the methods used and the problems involved in conducting research on a large scale, and to assist in a specific investigation. In this way they develop an appreciation for the broad background skills that researchers use and need.

The scientific curiosity of the cadets, nurtured in the research laboratory, has become the basis for an extended, serious, and enthusiastic research participation program for many members of the Cadet Wing.

CADET TRAINING

The life of an Air Force Academy cadet revolves around both academic studies and military training.

The basis of the cadets's four years at the Academy is preparation for commissioning as an Air Force officer. The cadet student body is the Cadet Wing and is organized like a military organization. A Cadet Wing commander, with the rank of cadet colonel, leads the Wing through the academic year extending from mid-August to June. Under his leadership are four other cadet colonels, each commanding a Cadet Group. Each group is composed of 10 squadrons of about 100 cadets each, and each squadron is commanded by a cadet officer. At each level, the commanders are assisted in the operation of their units by a staff of cadet officers and sergeants. The duties help prepare cadets for leadership roles in the Air Force.

To coincide with the military structure, cadets wear shoulder boards to indicate their class and their rank within the class. A freshman is a cadet fourth class; a sophomore is a cadet third class; junior, cadet second class; and senior, cadet first class. Each class is also known by its year of graduation, e.g., Class of 1976.

Immediately after arrival at the Air Force Academy, basic cadets receive military training during a seven-week program led by upperclassmen.

Cadets live in two dormitories—Vandenberg Hall, located on the north side of the Cadet Area, and Sijan Hall, on the Cadet Area's south side. Both dorms have a total of 2150 two-people rooms, some of which can be converted to accommodate three people. Each cadet has a bed, desk, lamp, chair, closet, wall bookcase, and chest of drawers.

Classes and study periods are from 7:30 A.M. to 3:30 P.M. on weekdays, with an hour break for lunch. Unless cadets are on varsity teams, they play on squadron intramural teams two afternoons a week following classes. The other three afternoons are set aside for extracurricular activities, studies, and drill or weapons firing. Saturday mornings are devoted primarily to military training, Cadet Wing parades, and inspections.

During the academic year, cadets are required to be in the Cadet Area after 7:45 P.M. Sunday through Thursday. These evenings are reserved for study. Taps is at 11:00 P.M., but cadets may stay up later

During their first summer new cadets participate in realistic survival exercises in the Rocky Mountains.

In initial summer training cadets—with limited rations and gear—sustain themselves while finding their way back to the Academy through the Rampart Range near Colorado Springs, Colo.

to complete their studies.

Cadets receive about $350 per month, in addition to their education, food, housing, and medical and dental care. They pay for their uniforms, books, clothing, and other personal needs. A portion of their pay is saved for them, and they receive the savings as a lump sum when they are graduated.

In July each year a cadet class of 1500 enters the Air Force Academy. Being admitted to the Academy represents an achievement for each individual. It means that the young man or woman was successful in obtaining a nomination and competing with many others

Hand-to-hand combat is also taught at the Air Force Academy.

seeking admission. Each phase of education and training he completes during the next four years represents another step toward the final goal of graduation with a Bachelor of Science degree and a commission in the Regular Air Force.

The first phase a new cadet undergoes is basic cadet training, which is the greatest transition he or she must make at the Academy. The training transforms each from a civilian to a military cadet, and lays the foundation upon which the Air Force officer of the future is built. The change from a relaxed civilian atmosphere to a disciplined environment can be difficult for some cadets. It requires learning to live by firm rules and to obey commands instantly.

The transition begins immediately. Upperclass cadets supervise three weeks of basic military instruction and intensive physical conditioning conducted in the cadet area. Basic cadets learn to march, to perform the manual of arms, and to participate in parade and review ceremonies. They undergo physical conditioning, beginning with basic exercises such as push-ups, knee bends, body twists, chinning, and running. Once toughened to physical exertion, they are tested by a rugged obstacle course. They participate in competitive sports, which are climaxed by Field Day events.

Basic cadets must run, jog, or doubletime everywhere. The pace of their daily routine is strenuous. Throughout the training, upperclassmen put the new cadets through many exercises of discipline to demonstrate how to live under constant pressure without disputing the reasons, as Air Force officers may be required to do.

At the completion of this indoctrination, basic cadets march to the Jack's Valley encampment on the Academy site where they live for three weeks under field conditions. They wear fatigues and combat boots, sleep on cots in tents they erect, and perform many challenging tasks to test their skills and endurance. Among these challenges are the confidence course, which is a series of obstacles designed to build the cadet's confidence in his physical ability; the armed and unarmed combatives course, which provides realistic training in self-defense;

"Sir!!" Size doesn't mean a thing as this cadet officer (right) braces a "Doolie" on the parade ground.

Basic cadets participate in Field Day activities at the end of basic cadet training. During the year all cadets must compete in either intercollegiate or intramural sports.

and the leadership reaction course, which utilizes team ingenuity to solve combat problem situations.

Basic cadets move at a fast pace from reveille at 6:00 A.M. to taps at 9:00 P.M. On week nights they are restricted to the cadet dormitory or site of field training. On weekends their recreational privileges are limited to scheduled functions. These generally include an outing at the cadet recreation area, a reception in the cadet social center, and dining out at the home of an Academy officer.

Basic cadets who successfully complete the summer training are accepted as members of the Air Force Cadet Wing at a parade late in August. Each one has undoubtedly recognized that he or she can withstand more strain and pressure than believed. They have developed confidence in their abilities to function individually and as a team in military situations, and they are eager to advance in their military training and begin their education as a fourth classman.

The Underclass Years

After the summer orientation to the Air Force, the new cadet is ready to begin the program of leadership development that will progress gradually over the next four years. It is designed to challenge the cadet to develop his or her intellectual, military, and physical capabilities to the maximum.

Intellectual preparation for leadership is the purpose of the academic program, conducted during fall and spring semesters for four years. The curriculum provides cadets with a wide range of understanding in the major areas of knowledge—the basic and engineering sciences, the social sciences, and the humanities. Each cadet completes a balanced sequence of core courses in those areas. The curriculum also provides for specialization with majors offered in twenty-three fields. During a cadet's fourth and third class years, he concentrates on core courses. These are basic studies that provide the foundation for upperclass courses that are professionally oriented toward Air Force careers.

Included in the core curriculum are courses in military studies and physical education. These are conducted during fall and spring semesters, as well as in summer training programs. During a cadet's fourth class year he attends a military studies class in the structure and combat capabilities of the United States and foreign defense forces. The third class summer includes three weeks of intensive SERE training (survival, evasion, resistance, and escape), conducted in Saylor Park in the nearby Rocky Mountains. Cadets learn how to make shelters, hunt for food, navigate by compasses and maps, and generally endure the hardships in a wilderness area. They also engage in simulated combat situations to practice techniques of survival and evasion in enemy territory. Finally, they are subjected to intensive prisoner-of-war training. Third class cadets also have a three-week diversified program of military training with several airmanship options available to them.

Physical education during the fourth and third class years is composed of basic fundamentals in aquatics, body development, combatives, and carry-over skills. Cadets are instructed in wrestling, boxing, and judo to develop physical confidence and learn skills of self-defense. In addition they receive instruction in individual sports, which they may continue to pursue after graduation, such as tennis, golf, squash, handball, and swimming. Each cadet participates in intramural or intercollegiate sports each semester. Competitive athletics develops courage, stamina, self-control, and teamwork.

Military and physical training are based on the concept that the heart of an Academy education comes from development of such ideals as tradition, honor, ethics, discipline, patriotism, courage, motivation, and dedication. Although cadets learn many military skills, they profit equally by acquiring the proper attitudes and precepts of leadership.

The laboratory for leadership training is the Air Force Cadet Wing. All cadets are members of the Wing, which is operated like an air base military organization. The Wing is divided into groups and squadrons with upperclassmen acting as officers and noncommissioned officers. Underclassmen are followers in the Wing, assuming that a cadet must learn to follow before he can understand and practice the duties of command.

In addition to functioning as a leadership laboratory, the Cadet Wing fosters pride and excellence through competition between the groups and squadrons. Cadet organizations compete in a variety of intramural athletics and in parades, reviews, and drill competitions.

The Upperclass Years

First and second class cadets advance to academic courses that relate to Air Force careers. Engineering courses provide preparation for aeronautical careers, and social sciences and humanities courses give the background for military leadership. Before their second class year begins, cadets must select a major from twenty-three subject offerings. Approximately one-half of the cadets have chosen a science or engineering major, and the other half have chosen a major in the social sciences or humanities.

The academic curriculum is further individualized through an enrichment program that challenges a cadet to work to the limits of his or her intellectual capacity. A variety of enrichment courses enables cadets to study in depth in their major or to broaden intellectual experiences in other subject areas of interest. Cadets may compete also for national scholarships or fellowships to obtain advanced degrees.

Military training for upperclassmen affords many opportunities to gain leadership and management experience. Upperclassmen hold officer and noncommissioned officer rank, which is identified by shoulder board stripes on their uniforms. They are in charge of summer training of the lower classes, as well as the operation of the Cadet Wing. They participate on many boards and committees that establish policies for the Wing.

Every upperclass cadet must spend one summer training period on Operation Third Lieutenant duty with an Air Force unit in the United States or overseas. This provides valuable firsthand experience in Air Force operations. Upperclass cadets have several optional summer training programs at the Academy and other military installations. The programs may vary from year-to-year, but the overall purpose remains the same: to place cadets in new and challenging Air Force environments to broaden their experiences.

Physical education and athletics for upperclassmen emphasize physical leadership training, advanced skills in combative situations, importance of physical fitness, and instruction in carry-over skills. Upperclassmen continue to participate in intramural sports, and they act as assistant coaches and officials in contests among the cadet squadrons. During a cadet's entire four years he must take a physical fitness test each semester. Many cadets from all classes participate in one or more of the nineteen intercollegiate sports. The opportunity for a number of cadets to compete is broadened by extensive schedules arranged at the freshman, junior varsity, and varsity levels.

Airmanship

A variety of airmanship programs is offered to cadets at the United States Air Force Academy under the direction of the Commandant of Cadets. The aviation curriculum is designed to educate, train and motivate cadets to be future Air Force leaders by providing them with the opportunity to gain practical leadership experience in an air-oriented environment.

A Schweitzer 2-22 sailplane, one of several types of long winged craft used in the soaring program for cadets, sails high over the Air Force Academy. All fourth classmen (freshmen) get up to six sailplane flights during their freshman year. Thereafter, selected volunteer cadets can progress through the program and earn private glider pilot, commercial glider pilot and glider pilot instructor ratings. Soaring is one of the Airmanship Division programs which also includes light plane flying, parachuting and ballooning.

Jet flight indoctrination is a part of the Airmanship program.

Airmanship programs include jet orientation flights, light aircraft training, soaring, parachuting, parasail training, and balloon operations. A cadet can advance to become an instructor in powered aircraft, sailplanes, or hot-air balloons, or qualify as an instructor parachutist.

All cadets, whether they meet Air Force pilot qualifications or not, and regardless of background, can and do participate in a variety of flying programs. For the medically-qualified individuals, the courses are oriented to motivate a cadet toward a rated career in the Air Force. For other cadets, the airmanship programs are designed to provide nonrated officers with a working knowledge of air operations to enhance their insight into the overall Air Force mission.

This aviation curriculum is an essential and major element of the total academic, military, and athletic environment of the Air Force Academy. The success of these programs is reflected in the professional performance of Academy graduates.

A senior cadet with an instructor pilot flies over the Air Force Academy in a T-41. The program is highly standardized to mentally prepare the cadet for rigorous Air Force pilot training.

Cadets involved in the Academy's parachute program land on the cadet parade field.

The cadets' first contact with airmanship occurs during Basic Cadet Training shortly after arriving at the Academy. Each new cadet receives a thirty-minute orientation flight in a jet trainer. This program, known as "Stardust," provides the young people with knowledge of a jet's flying characteristics. They gain additional exposure to jet flying through Airmanship (AM) 370, which gives cadets an appreciation of jet aircraft operations and aircrew responsibilities, while providing a practical application laboratory for flight-oriented academic courses.

84

AVIATION SCIENCE DIVISION

The Aviation Science Division, under the direction of the Commandant of Cadets, offers a variety of aviation curriculum courses to the Cadet Wing. These flying courses range from an introduction to flying through basic aviation skills and techniques to advanced navigation ability and checkout as a cadet instructor. The basic course, Aviation 460 (Studies in Aviation Fundamentals) is designed for the cadet who will pursue nonrated duties upon graduation. The advanced flying courses are designed specifically to better prepare cadets for entrance into Undergraduate Pilot and Navigator Training. Additional courses include an avionics systems course, Aviation 490, which looks at current and future navigation systems in terms of functional analysis and operational capabilities.

The primary aircraft used throughout the various courses offered by the Aviation Science Division is the T-43A. Designed and manufactured by the Boeing Company, the basic T-43 design and construction are derived from the Boeing 737. It is a swept wing, twin engine jet which flies at altitudes up to 35,000 feet and airspeeds up to approximately 450 knots.

The Aviation Science Division's curriculum does not confine cadet study to atmospheric flight. Courses in astronomy offer exposure to space and space related flight in descriptive astronomy and applied astronomy. Both courses extensively use the Academy planetarium and observatory and offer realistic flight laboratory challenges using the T-43 aircraft. With the astronomy and avionics courses added to the flying and trainer experiences, the Division affords cadets the opportunity to study aviation to any level of expertise. For the pilot and navigator bound cadet, the advantages are obvious. For the nonrated graduate, the Aviation Science Division courses offer one of the few chances to experience the flight environment as a crew member and better understand the primary mission of the Air Force.

Aviation Club   The Cadet Aviation Club utilizes three Cessna 172s, two Grumman-American Travelers, a Beechcraft Sierra, and a Beechcraft Sundowner provided by the Academy Aero Club. The cadet must pay aircraft rental fees and instructional costs, but is partially reimbursed from Cadet Welfare Funds. Membership in the Aviation Club is normally around 105 cadets. In addition to flight and ground instruction, cadets participate in field trips to various Air Force and civilian installations to participate in fly-ins and aerial competitions, and to tour aeronautical facilities. The Academy Aviation Club is a member of the National Intercollegiate Flying Association.

Soaring   Other airmanship courses offered to cadets are those in the soaring program, which is one of the largest and most active in the United States. The Academy's location affords excellent conditions for mountain wave and thermal soaring, enabling cadets to reach altitudes in excess of 25,000 feet. Sailplane training is designed as an air-oriented leadership opportunity for cadets as well as a highly motivational program. These aspects are enhanced by the fact that eighty-five percent of the training is conducted by highly qualified cadet instructor pilots. The Academy sailplane fleet consists of fifteen Air Force-owned Schweitzer one-and two-place sailplanes and one high performance Phoebus "C" sailplane. The soaring program currently selects and trains 375 cadets each year, all of whom receive at least one sailplane orientation flight during their first year at the Academy. During the following years, participation is voluntary, and soaring cadets progress at a standard rate.

T-41 Pilot Indoctrination Program   The largest and most important cadet flying program is the T-41 pilot indoctrination program, which began in the spring of 1968. The flight training for this program is conducted by Air Training Command at the Academy airstrip. Ground school instruction is provided by Airmanship instructors and held in the academic area.

This program is required for all physically qualified first classmen (seniors) who plan to enter pilot training after graduation. The program is highly standardized to prepare the cadet mentally for rigorous Undergraduate Pilot Training (UPT).

Parachuting   Cadet parachuting at the Academy began in 1963 when a recreational club was formed by interested cadets. The program was carried out at the Academy airstrip. In 1966, the club status was discontinued when the Air Force approved the Academy's basic free fall parachute course.

Cadets have been attending the Army parachute training program (airborne) at Fort Benning, Georgia At present approximately 500 cadets participate in this challenging program as one part of their summer training. The majority of them are third classmen (sophomores), but approximately 100 first and second classmen (seniors and juniors) also attend. Airborne training involves mass static line jumping from aircraft flying at a relatively low altitude. This training exposes cadets to the operations and personnel of the U.S. Army, which they may be supporting in their future Air Force careers through troop carrier or close air support operations. Successful completion of airborne results in the award of a coveted Parachute Badge.

The basic Academy course is built on the unique parachute requirements of aircrew members. Primarily they must be skilled in free fall parachuting techniques for emergency bailouts from disabled aircraft, often at high altitude. They must also master the use of steerable parachutes and the basic flying equipment worn by aircrew members.

Three parachute training programs currently exist in the cadets' airmanship programs—the basic, advanced, and instructor training courses. A fourth course gives military training credit to cadets who are instructors for the other three programs. They play a major role in training and safety. For example, virtually 100 percent of all cadet jumps are supervised by a cadet jumpmaster.

These duties provide practical leadership experience with a large amount of responsibility. This responsibility includes ground training, equipment fitting, preflight inspection, aircraft loading, all prejump actions to include emergencies, and safe and orderly egress from the aircraft. The regular Air Force training staff, consisting of two officers and three noncommissioned officers, limit their involvement to that of advising and supervising the cadet instructor.

Most of the jump training is conducted on the Academy airstrip using two U-4B aircraft carrying up to six jumpers each.

Parasailing Parasailing introduces cadets to the basic principles of parachuting. Cadets are taught the fundamentals of canopy control and parachute landing techniques. A parasail is a parachute canopy designed to produce lift as it is towed on a 600-foot rope behind a moving vehicle. The cadet is towed to an altitude of approximately 300 feet and then lowered gently to the ground by decreasing the speed of the vehicle. All cadets receive one or more orientation tows during basic cadet training, and they may volunteer to participate on weekends during the academic year. The parasail program is organized and operated by cadet instructors, providing another opportunity for practical leadership training in a flight-related activity.

Ballooning The Airmanship Division also operates another extremely unusual and interesting activity, the Cadet Hot Air Balloon Club. Ballooning is strictly voluntary, and provides the cadet with the opportunity to experience aviation as the early pioneers did. The program provides insight into the relationship between classroom aerodynamics and the actual flight environment. This activity, in addition to being recreational and motivational, teaches cadets the fundamentals of air currents, weather, and weather patterns. Professionalism, leadership, safety, and development of a sense of responsibility are stressed throughout the program.

Athletic Program

Not all athletes become Air Force Academy cadets, but all Air Force Academy cadets become athletes. Few schools in the country have as broad or as extensive physical education, intramural, or intercollegiate programs.

The program develops desirable traits such as persistence, the will to win, aggressiveness, and courage—characteristics that are attributes of a good leader. Positive attitudes toward physical fitness and esprit de corps of the Cadet Wing are added goals.

Each cadet takes physical education courses throughout the four years at the Academy. During the fourth class (freshman) year, instruction is primarily designed to develop physical fitness and basic combative skills. The courses include swimming and boxing for men, physical development and fencing for women, and coeducational classes in swimming and gymnastics.

A full range of varsity sports—including basketball—is available to cadets at the Air Force Academy.

The popularity of football at the Academy is reflected here with a touchdown against Navy and the crowd is on its feet.

During the third class (sophomore) year, cadets take tennis, squash, handball or racquetball plus a physical fitness methods course which will enable them as officers to maintain fitness throughout their service career.

Courses during the second class (junior) year include judo, volleyball, survival swimming and golf. Unarmed combatives, badminton and life saving are offered to first class cadets (seniors). First classmen are also allowed to elect advanced instruction in two sports of their choice.

Cadets of Air Force Academy cheer as their varsity team scores a touchdown against West Point.

The intramural program, consisting of 16 sports, is conducted by the cadets under the supervision of officers from the Department of Physical Education. The program gives upperclassmen experience in organizing, coaching, and officiating. Each cadet squadron is required to have a team in each sport each season.

Nineteen intercollegiate sports are played and represent the highest form of competition for cadet athletes. Varsity teams compete in football, cross country, soccer, water polo and volleyball during the fall season. During the winter sports season cadets compete in basketball, fencing, ice hockey, gymnastics, indoor track, wrestling, swimming, rifle and pistol. When spring comes varsity teams in baseball, outdoor track, lacrosse, golf and tennis carry the banner of the Air Force Academy.

Cadets enjoy the best athletic facilities available anywhere. The cadet gymnasium contains squash courts, handball courts, basketball courts, two swimming pool areas, an indoor rifle and pistol range, gymnastic, wrestling, fencing and boxing rooms, a weight room and, of course, lockers.

The Field House contains a ⅙ mile indoor track with an Astro-turf infield, a basketball court with Tartan floor for varsity competition

The prairie falcon, native to Colorado, is the falcon seen most by the public. An exhibition is given at each varsity football game, at home and away. Usually at halftime the falcon is released to the center of the field and the cadet falconer swings a lure as the falcon dives again and again, attempting to catch it. The falconer finally tosses the lure in the air, and the bird, more often than not, catches it and settles to the ground. Young prairie falcons are captured in their natural habitats in the vicinity of Cathedral Rock north of the academic area. They are raised and trained at the Academy by volunteer cadet handlers.

and a regulation ice hockey arena.

In addition to Falcon Stadium which seats 50,000 for football games, outdoor facilities include 45 multi-sports fields for intercollegiate and intramural competition covering 125 acres, plus two baseball diamonds and an all-weather track.

Naturally, not all is athletics, studying, and military training at the Air Academy. Many organized extracurricular activities are available so cadets may develop their creative talents and hobbies, as well as their professional interests. Cadets organize and continue the various groups on a voluntary participation basis.

There are forty-five clubs or groups ranging from singing to scuba diving, from parachuting to photography.

Typical representative activities are the Forensic Association, Cadet Forum, Mathematics Club, Water Polo Team, Cadet Bowling Team, Parachute Team, Professional Studies Group, and Special Warfare Group.

The publications area includes *Contrails,* the fourthclassman's handbook; *Talon,* the cadet monthly magazine; and *Polaris,* the cadet yearbook.

Clubs that provide recreation and pursuit of various hobbies include the Aviation Club, Bowman Club, Gun Club, Mountaineering Club, Saddle Club, Gavel Club, and a Music Club that supplies cadet bands for cadet dances and parties.

Each organized cadet activity contributes to the training of the cadet by giving him an opportunity to accept responsibility and to exercise authority in managing and directing it.

Participation in organized extracurricular activities contributes to developing the potential of the future officer while providing him or her constructive areas of pursuit for his or her free time.

Graduation

Graduation marks the end of one challenge and the beginning of another. It is the goal of the four-year development toward a degree and a commission in the Air Force. The years haven't been easy. They have been filled with assignments that challenged the depths of each man's mental, physical, and moral capacity. By comparison, the Air Force Academy is much tougher than most civilian universities. The Academy requires for graduation a minimum of 180½ semester hours with at least a C average in academic, military, and physical education courses. It also requires each cadet to demonstrate an aptitude for commissioned service and leadership, including conduct and demeanor worthy of the rank he will hold. Fulfilling these high standards is a rewarding feeling for the graduate. It gives him confidence in his abilities and makes him proud of his achievements. It gives him a sense of realization that worthwhile goals in life do not often come easy, but in the long run the rewards are worth the efforts.

The Academy has provided each graduate with one of the finest educations available in the country. The Academy's outstanding curriculum, faculty, and facilities have all contributed to this effort.

A comprehensive schedule of social activities—including dances and parties—are part of cadet life at the Air Force Academy.

Cap tossing at the Air Force Academy takes place at Graduation ceremonies following the graduates' last order from the Commandant of Cadets, "Gentlemen, you are dismissed." In a burst of enthusiasm, the new second lieutenants cheer and loft their dress caps high in the air. Their cadet days are over, and now they begin their careers as air force officers.

The academic program and the complementing leadership training have provided each graduate with the essential elements for successful service as an officer in the United States Air Force.

Questions and Answers About the Air Force Academy

What is the overall mission or purpose of the Air Force Academy?

The mission of the Academy is to provide instruction and experience to each cadet so that he or she graduates with the knowledge essential to leadership, and the motivation to become a career officer in the United States Air Force.

How large is the Air Force Academy site?

Approximately 18,000 acres.

When was the Academy established?

The bill authorizing the Air Force Academy was signed by President Eisenhower on April 1, 1954. The first class entered in July 1955 at interim facilities on Lowry AFB, Denver. The Cadet Wing moved into its permanent facilities near Colorado Springs in late August 1958. The first class was graduated in June 1959.

U.S. Air Force Academy cadets dine in modern Mitchell Hall at USAF Academy.

Cadets walking near the statue of Pegasus reflected in the windows of Arnold Hall at the U.S. Air Force Academy.

Vandenberg Hall on the left and Fairchild Hall with the fountains of the mall in the foreground at U.S. Air Force Academy.

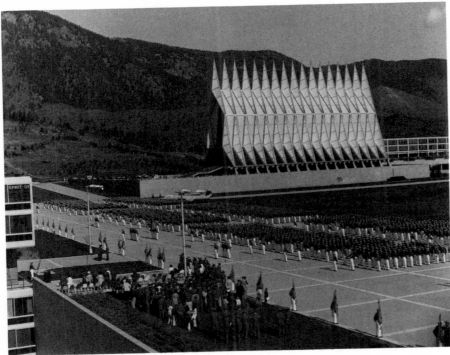

The 4,000 plus cadets assemble on the Academy's terrazzo during dedication ceremonies of the Sijan Hall dormitory.

Where do the cadets live?

They are housed in two cadet dormitories. One of these dormitories, Vandenberg Hall, is the largest building at the Academy, ¼ of a mile long, 6 stories high, and containing 1320 cadet rooms, squadron meeting rooms, hobby shops, a cadet store, and barber shop. It was named for General Hoyt S. Vandenberg, former Air Force Chief of Staff. The other dormitory, Sijan Hall, has similar facilities, but also includes a tailor shop and dental clinic.

How was Colorado picked as the site of the Air Force Academy?

A site selection commission was appointed by then Secretary of the Air Force Talbott in 1954. After traveling over 20,000 miles and visiting 22 states, the commission recommended three locations of the many proposed: Colorado Springs, Colorado; Alton, Illinois; and Lake Geneva, Wisconsin. From them Secretary Talbott selected this site north of Colorado Springs.

Where do the staff and faculty live?

There are two housing valleys in the center of the Academy called Pine and Douglass Valleys. In the Valleys are 1200 privately-constructed Capehart housing units for officers, NCOs and their families. Occupants of these houses give up their normal rental allowances to pay off the mortgages and provide maintenance. In addition, forty-three sets of senior officers' quarters were built with appropriated funds. Academy families also occupy eighteen houses that were on the land when purchased. Some of the faculty and staff live off base.

Why is the Academy spread out so much?

The Academy is intentionally divided into four major areas to separate cadet, support, and family functions. The northern part is the cadet area. The center portion contains the two housing valleys and the Community Center, which provides the recreational and shopping facilities for assigned military personnel. It also includes the airmen's dormitory, dining hall, gym, and the Academy Preparatory School. The third section, located to the south, contains the air police headquarters, motor pool, NCO Club, photo lab, warehouses, and supply, engineering, and maintenance facilities. A fourth portion, an airstrip, is in the southeast portion, adjacent to U.S. Highway 85-87.

Why was the falcon selected as the mascot of the Academy?

The falcon has characteristics that best typify the U.S. Air Force; speed, beauty, graceful flight, courage, keen eyesight, alertness, noble carriage, and noble tradition.

How many falcons do you have here?

The cadets usually keep about a dozen falcons in the mews near the gym. One or two usually perform at Air Force Academy football games.

What are the major buildings in the main Cadet area?

Mitchell Hall is the cadet dining hall, which can accommodate the entire cadet wing at one sitting, as it contains almost two acres of unobstructed floor space. It is named after General Billy Mitchell, pioneer of military aviation. *Fairchild Hall,* the academic building named for General Muir S. Fairchild, a pioneer in Air Force education, is six stories high and divided into two sections. The cadet library, dispensary, and offices of the Commandant of Cadets' staff occupy most of the northern section. The larger section is the classroom complex. The sixth floor contains offices of the faculty. The fourth and fifth floor are devoted to classrooms, while the third floor consists of auditoriums. The second floor contains laboratories. *Cadet Gym* contains two swimming pools, basketball and volleyball courts, boxing and wrestling rooms, squash courts, handball courts, a gymnastics room, and a rifle and pistol range. Surrounding the gymnasium are a number of athletic courts and fields. *Arnold Hall* is the Cadet Social Center, named after General Henry H. "Hap" Arnold, World War II Air Force leader. It houses a ballroom, theater, recreation rooms, and a cadet snack bar. *Planetarium,* a dome-shaped structure, which contains a Spitz projector to display the heavens at any point in time, is used to teach celestial navigation and astronomy. Public showings are scheduled. *Harmon Hall,* named after Lt. Gen. Hubert R. Harmon, the first superintendent of the Air Force Academy, is the administration building, housing the offices of the superintendent and his staff. *Field House* is a multipurpose area that includes a 6600-seat basketball arena, a 2400-seat hockey arena, and office facilities for the Department of Athletics.

How many people are at the Air Force Academy?

The Academy has approximately 1,100 officers assigned, about 550 of whom are on the faculty; approximately 1,500 enlisted personnel; about 2,100 civilian employees; and the cadet strength, which is approximately 4,100. Total strength is about 7,150. In addition there are about 4,000 dependent wives and children living on the grounds.

At what times does a cadet get up?

The cadet's day starts at 06:15 in the morning and ends at 10:45 P.M. If a cadet wishes to study beyond 10:45 P.M. he may do so.

When does he go to classes?

The average cadet has 4½ hours of classroom work per day. There are four class periods in the morning and two in the afternoon.

When does he get his military training?

It is accomplished throughout the entire week with Saturday mornings devoted in part to military training, as are the summer months from the beginning of June through the third week in August.

When does a cadet participate in athletics?

Intramural athletics and intercollegiate practice are scheduled in the afternoons. Formal physical education classes are scheduled throughout the day.

How many cadets do you have?

During the academic year, there are more than 4,000 cadets enrolled at any given time. Total authorized strength of the Cadet Wing is 4,417.

What increase of facilities will be required to accommodate the additional cadets authorized at the Academy?

A new dormitory, a new field house, an expansion of the dining hall, academic building, gymnasium and social center.

What is the pay of a cadet?

A cadet is paid one-half the base pay of a second lieutenant. At present this is about $350 a month.

What is this money used for?

From this allowance a cadet must pay for his distinctive cadet uniforms, many of his textbooks, dry cleaning, laundry, haircuts, and personal items.

What does the Government provide in addition to a cadet's pay?

The Government provides tuition, room, board, medical and dental care.

What do the shoulder-boards designate?

They designate the cadet's class and rank.

What is the military organization of the body of cadets?

The body of cadets is designated the "Cadet Wing." It is composed of 40 squadrons of approximately 100 men each and is divided into four groups of 10 squadrons each.

How are cadets assigned to the squadron?

Each cadet squadron is made up of members of all four classes. First classmen (seniors) hold the command ranks, and are designated cadet officers. Second classmen (juniors) hold the noncommissioned officer positions. Third and fourth classmen hold no rank, except for a few third classmen who hold the rank of staff sergeant.

What are the freshmen called?

Fourth classmen (freshmen) are called "doolies," a slang derivation of the Greek word *doulos,* meaning slave.

Can a cadet marry while at the Academy?

To be appointed a cadet a young person must be unmarried, and agree not to marry prior to graduation.

Are there any Black cadets?

The Air Force Academy entered its first three Black cadets in 1959. Now more than ten percent of the cadets are from racial and ethnic minorities.

Are cadets required to attend chapel service?

No.

May cadets have automobiles?

A cadet may own an automobile in his first class (senior) year.

Can cadets leave the Academy grounds?

Privileges to leave the Academy are based on gradual transition from the status of a new cadet to that of a second lieutenant. The new cadet, a fourth classman, is very restricted in the number of privileges he may receive. He is permitted to leave the Academy only on specific holidays, after certain varsity football games, for special cadet activities, and for occasional dining with close relatives or staff officers. A first classman, however, is relatively free during off-duty hours, just as he will be as a second lieutenant. Thus, privileges are progressively increased by class in recognition of added maturity and responsibility.

Do the cadets run their own squadrons?

There is a parallel chain of command, one made up of cadets, the other of commissioned officers. The cadets, for all intents and purposes, administer the Wing, the officers acting in supervisory and advisory capacities. The commissioned officers directly involved in the squadrons and groups are known as Air Officers Commanding. The Commandant of Cadets is by law the Commander of the Cadet Wing.

What is the Honor Code?

The Honor Code states: "We will not lie, steal, or cheat, nor tolerate among us anyone who does." The cadets administer the honor system themselves, and are intensely proud of it.

Do you have officers from other countries assigned to the Academy?

There are several officers of other countries assigned to the Academy faculty and staff on an exchange basis. Included are officers from France, Germany, Latin America, Great Britain, and Nationalist China.

What extracurricular activities do the cadets have?

There are approximately fifty extracurricular cadet activities ranging from hunting and fishing, skiing, and mountain climbing clubs to debating, to membership on the staffs of the cadet magazine and yearbook.

When can we see the cadets march?

They march in formation to most meals. Formal parades are held frequently on Saturday mornings and during June Week.

Are the cadets well fed?

Menus at the Cadet Dining Hall are based on 4500 calories per day. Actual consumption runs nearer 5300 calories.

Is the Academy Band composed of cadets?

No. The Air Force Academy Band is a regular Air Force Band made up of noncommissioned officers and airmen.

How do young men entering the Air Force Academy compare with those entering other colleges and universities?

In terms of high school grades and scores on nationally used college entrance tests, the average Air Force Academy cadet falls in the top fifteen percent of all young men admitted to colleges and universities in recent years.

How much is included in the total curriculum?

The prescribed curriculum totals 180½–187½ semester hours. This includes 138–144 semester hours in academic subjects, 28 hours in military training, and 14 semester hours in physical education. (One semester hour of credit is awarded for completion of 35 hours of study. Normally, this includes 4 classroom attendances and 21 hours of outside preparation.)

How heavy is the cadet's academic workload?

The minimum prescribed academic program totals 138 semester hours. The core curriculum, about 111 semester hours, is balanced almost evenly between basic and applied sciences and the humanities and social sciences. Elective courses, totaling about 27–33 semester hours, are selected by each cadet to complete a major.

Do the cadets study about rockets and missiles?

Yes. In 1957 the Academy established the first undergraduate Department of Astronautics in the nation, and today every cadet is required to complete at least one prescribed course in astronautics.

What foreign languages are required?

Each cadet is required to complete at least 4½ semester hours of language. Arabic, German, Russian, French, Chinese , Japanese and Spanish are the regular languages offered.

How often do cadets have to take exams?

It varies, depending upon the subject concerned, but the average is one evaluation for each three to four classroom sessions. They vary from quizzes lasting five to ten minutes to four-hour finals.

How much must cadets study?

In most courses, a cadet is expected to study one and one-half hours for each hour in the classroom.

Do you have any civilian instructors?

The Air Force Academy has two civil servants from the Dept. of State and three distinguished visiting professors on the faculty. Part of the mission of the Academy is to motivate cadets toward a career in the service. It is our opinion that this can best be done by military officers. Each officer assigned to the faculty holds at least a master's degree in the field in which he is teaching, and has some experience in that field. Roughly, twenty-six percent of the faculty members hold a Ph.D.

What is the tenure of the faculty?

The Air Force Academy is authorized twenty-one permanent professors, plus the Dean of the Faculty, and a number of tenured associate professors who normally are expected to remain at the Academy throughout their military careers. Other faculty members generally serve as Academy instructors for four years.

What academic average is required for a cadet to graduate?

To graduate a cadet must have an overall academic average of 2.0, which is a C; 4.0 is an A; 3.0 a B; and 1.0 is a D. Additionally, he must have a 2.0 average in the courses selected for his major.

What happens to cadets who fail?

A board of officers meets at the end of each semester to determine the future of any cadet who has failed a course or failed to maintain a semester or cumulative grade point average of at least 2.0. At the board's discretion, the cadet may be allowed to go ahead with his class, but repeat a failed course or be turned back to join the next lower class, but not repeat subjects successfully completed. Finally, a cadet who simply does not have the ability to get through may be discharged, returning to civilian life or to the military, depending on his status prior to entering the Academy.

Can Air Force Academy cadets transfer to another service on graduation?

Yes, a limited number can transfer to another service upon approval of an application to do so.

What do cadets receive at graduation?

They receive an accredited Bachelor of Science degree, with a major in an academic subject area, and a regular commission as a second lieutenant in the U.S. Air Force.

How many Rhodes scholars has the Academy had?

The Air Force Academy has had 19 Rhodes scholars. From 1959 through 1967, a total of 281 distinguished scholarships and fellowships have been won by Academy graduates in national competition.

Do all cadets in each class take the same courses?

No. Twenty-three undergraduate majors are available, ranging from history to astronautics. Under the Air Force Academy's curriculum-enrichment program, cadets may earn credit for courses at an accelerated rate, and by taking overload courses, exceeding the normal semester-hour load. Thus, there is much variety in the curriculum for members of each class.

How large are the academic classes?

The average class section consists of fifteen to twenty students. This allows the instructor to establish a rapport with each individual cadet and to recognize better each student's strengths and weaknesses. The Academy leans toward the seminar approach to instruction, and lectures are held to a minimum.

How are cadets assigned to academic classes?

Most sections are arranged homogeneously, with every man in each section the intellectual equal of the others. Most of the details of the complex scheduling problem are handled by sophisticated computer programs.

What kind of subjects do the cadets study in their military training?

The military training program, which qualifies each graduate to be commissioned as a regular officer in the U.S. Air Force, distinguishes the Academy from other universities. In addition to the professional subjects taught in the military studies classroom and during field study, portions of each day are devoted to a program of practical leadership and command.

What do the cadets do in the summer?

Cadets spend their first summer at the Academy undergoing intensive military training and physical conditioning. Succeeding summers are spent on field trips, on home leave, and at the Academy as part of the first class detail that trains the new class.

What do cadets do on the U.S. field trips?

At the end of their first academic year, cadets take part in a field study of U.S. Air Force installations throughout the country. This gives them the chance to observe first hand and learn more about the U.S.

armed forces that they have studied the preceding year.

Do cadets learn to fly while at the Academy?

Although cadets do not complete flying training or receive their pilot rating at the Academy, those cadets that are programmed to go to pilot training do take their first phase of training while at the Academy. This is accomplished in T-41 aircraft under the supervision of Air Training Command personnel. It consists of thirty-six and a half hours of flying and thirty hours of ground school. Besides this, cadets also receive jet orientation flights in T-33 aircraft, and have the option of flying sailplanes, light aircraft, and taking navigation orientation courses.

What is the cadets' service commitment following graduation?

Five years.

What does the athletic program at the Air Force Academy involve?

Cadets participate in an extensive athletic program, composed of physical education courses, intramural sports, and intercollegiate athletics.

How extensive is the physical education program?

The program extends across the entire four years. In the first two years cadets learn, and participate in, the contact sports, such as football, boxing, and wrestling. During the last two years they become proficient in carry-over sports, such as golf, tennis, handball, and squash, which they may play throughout a lifetime to maintain physical fitness.

How does the intramural sports program operate?

Each cadet squadron fields teams and competes with other squadrons in sixteen intramural sports. The cadets also administer and officiate in the competition.

How extensive is the Academy's intercollegiate sports program?

Nineteen sports are included in intercollegiate competition. The Academy is not a member of a regional conference, but plays a nationwide schedule.

Do members of intercollegiate teams get any special treatment academically?

Athletes must take the same courses and maintain the same academic standards as other cadets. If an athlete is found deficient, he is dropped from intercollegiate participation.

How is your intercollegiate program financed?

Although the Air Force participates in all major sports, only three produce revenue—intercollegiate football, basketball and hockey. These funds are administered by the Air Force Academy Athletic Association, a nonprofit, nongovernmental organization. These support the financing of all intercollegiate sports.

Does the Air Force Academy play West Point and Annapolis in football?

Yes. Games with each have been played in the past. It is planned that in the future, games against one or the other will be scheduled each year.

What did the field house cost?

$6.5 million.

Are all cadets appointed by congressmen?

No. There are several categories of nomination, and a cadet may compete in one or more categories. Congressional appointments comprise approximately eighty-five percent of the total, but in addition there are presidential appointments, traditionally reserved for the sons of regular or reserve military personnel; vice-presidential appointments; appointments for sons of deceased veterans and sons of Congressional Medal of Honor winners; appointments for members of the honor military schools, members of the regular and reserve forces; and others.

What are the age requirements for entry?

A young man must be at least seventeen and must not have passed his twenty-second birthday on July 1 of the year he reports to the Academy.

What are the physical requirements?

A cadet must be in good physical condition, be between 5'6" and 6'8" tall, have all dental defects corrected, and pass a rigorous physical aptitude exam.

Do all Academy Cadets have to qualify for pilot training?

Approximately seventy-fivpercent of the cadets admitted to the Academy qualify for pilot training. The others are admitted under a waiver by the Academy. A waiver may be granted to a candidate who does not meet the vision or height requirements, provided his records indicate outstanding academic or leadership aptitude and achievement.

How may a young man prepare for the Air Force Academy?

He should prepare thoroughly in high school by taking four units of mathematics and four units of English, in addition to courses in general that cover the sciences, social sciences, and humanities.

Also, he should participate extensively in high school extracurricular activities, both athletic and nonathletic.

What tests must a cadet take to get into the Academy?

The College Entrance Examination Board tests or the American College Testing Program tests, a physical aptitude test; and a medical examination.

What should a young person who is interested in the Academy do first?

He or she should contact the Academy Liaison Officer in his community or write to the Registrar, United States Air Force Academy, Colorado, 80840, for information as to requirements and procedures of application. If his or her high school counselor does not know the liaison officer, the young person should contact the liaison officer coordinator in his state or local area. These names are found in the Academy catalog.

How many applicants do you have each year?

In the final competition, around 9,000 compete for entry in each new class.

Of the 9,000 (approximate) that compete, how many actually enter the Academy?

About one out of every seven.

4
The United States Coast Guard Academy

Because the U.S. Coast Guard is a small organization—with 44,000 officers and men it is the smallest of the Armed Forces—it places a premium upon leadership. Relatively early in their careers Coast Guard officers are called upon to assume command responsibility in one of the many units into which the service is divided. Preparing them for this responsibility, and for added responsibilities in later years, is the function of the Coast Guard Academy.

Aerial view of the U.S. Coast Guard Academy at New London, Connecticut.

HISTORY

Many of the Coast Guard's first officers learned their trade in the Continental Navy. When the Navy was disbanded after the Revolutionary War, these officers took their ability to handle men and ships

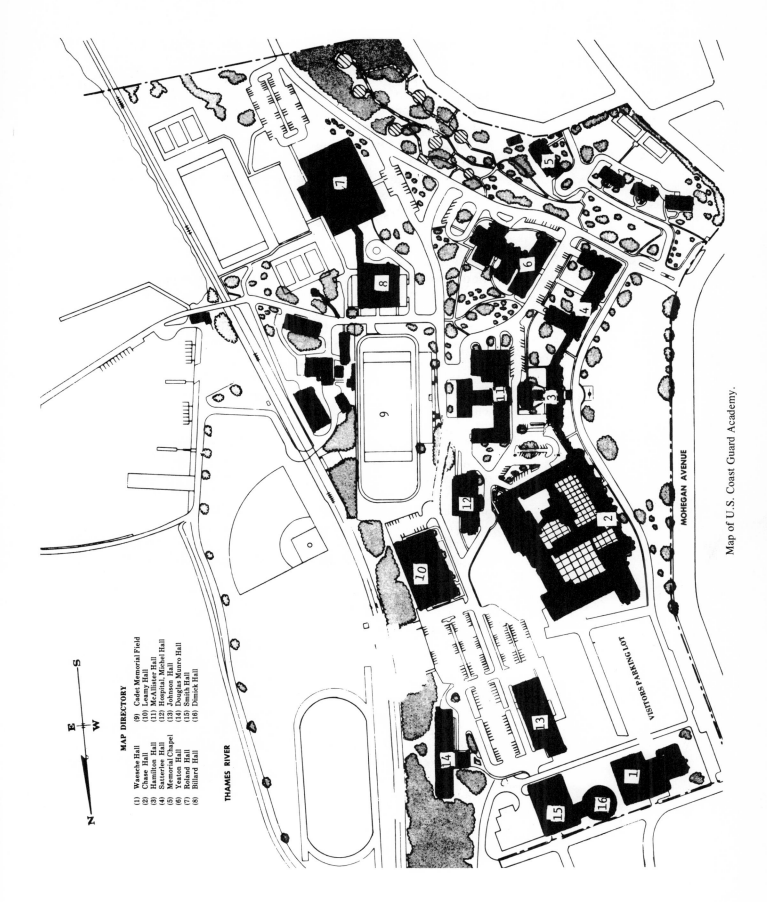

MAP DIRECTORY

(1) Waesche Hall
(2) Chase Hall
(3) Hamilton Hall
(4) Satterlee Hall
(5) Memorial Chapel
(6) Yeaton Hall
(7) Roland Hall
(8) Billard Hall

(9) Cadet Memorial Field
(10) Leamy Hall
(11) McAllister Hall
(12) Hospital, Michel Hall
(13) Johnson Hall
(14) Douglas Munro Hall
(15) Smith Hall
(16) Dimick Hall

THAMES RIVER

MOHEGAN AVENUE

VISITORS PARKING LOT

Map of U.S. Coast Guard Academy.

105

to Alexander Hamilton's cutter fleet, then known as the Revenue Marine. Other early Revenue Marine officers came from the merchant marine. Politics played a big part in appointment and advancement in those days. Once appointed, an officer retained his assignment and received promotions as long as he remained in favor with the local collector of customs, a situation that invited abuse.

Morale was low in the Revenue Marine when Louis McLane became Secretary of the Treasury in 1831. After an investigation of the organization's personnel problems, he announced a new policy: "With a view to greater efficiency in the cutter service in future, vacancies will be filled by promotions from among the officers in that service, when that shall be found preferable to other appointments, having regards to fitness as well as seniority."

McLane's stand helped matters somewhat, and reforms were continued by other Secretaries of the Treasury, notably John C. Spencer, who, in 1843, set up the Revenue Marine Bureau in Washington to centralize control of the service. He also instituted a system of original appointments only in the grade of third lieutenant and promotion by examination before a board of officers. In 1845, Congress, in the act establishing an engineer corps for the Revenue Marine, included the provision that no one "be appointed to the office of Captain, first, second, and third lieutenant of any revenue cutter, who does not adduce competent proof of proficiency and skill in seamanship and navigation."

More than thirty years passed before Congress moved to establish a training program for young men who wished to become officers in the Revenue Cutter Service. In the meantime, the Revenue Marine Bureau had ceased to exist, and politically inspired appointments and promotions again became the rule. The demands of the Civil War reversed this trend, however, and by 1865 a movement to reinstate centralized control of the Cutter Service was underway. Congress formally established a Revenue Marine Division in the Treasury Department in 1875.

Beginning in 1869, the Revenue Cutter Service took steps to remove undesirable officers from its rolls, many of them men acquired during the service's Civil War expansion. Seven of 19 captains and 33 of 103 lieutenants were released. Those that remained were awarded a rank determined by their qualifications. Moreover, regulations were adopted to eliminate political influence in appointments, promotions, and assignments. Appointments could be made only in the grade of third lieutenant or second assistant engineer, and candidates were required to have a certain amount of practical experience, and to pass physical and professional examinations.

The next step was a school in which the Revenue Cutter Service could train its own officer candidates. At the request of the service, Congress authorized a two-year cadet training program in 1876.

In December of 1876, nineteen young men between the ages of eighteen and twenty-five appeared before an examining board to demonstrate their knowledge of arithmetic, geography, and English. In addition, the candidates were graded on their "general aptitude."

Nine failed the written examination, and a tenth was failed in general aptitude. According to the examining board, he made an unfavorable impression "by his general deportment and by his manifest disposition to prevaricate."

Because experience aboard a sailing vessel was deemed the best way to prepare a young man for service at sea, the Revenue Cutter Service's first class of cadets reported to the schooner *Dobbin* in May 1877. The *Dobbin* had been fitted out as a schoolship under the command of Captain J. A. Henriques, one of the service's most able officers.

During each of their two years at the School of Instruction, cadets were to spend one term at sea and two terms ashore. While at sea they received practical training in seamanship, navigation, signals, and "exercises aloft." After the *Dobbin* tied up at New Bedford, Massachusetts, in October, the cadets began academic studies that included algebra, geometry, trigonometry, philosophy, astronomy, history, French, English, law, customs law, navigation law, and international law. A drill loft and a rigging shed served as classrooms.

Before the class of 1879 graduated, the *Dobbin* was replaced by the *Salmon P. Chase,* a bark-rigged vessel that had been especially designed as a schoolship. The *Chase* had accommodations for twelve cadets and, like the *Dobbin,* she tied up at New Bedford each winter while the cadets pursued their studies ashore.

Schooner James C. Dobbin, first practice ship of the first class of the Coast Guard Academy. She served as a school from 1876–78.

Early training ship Revenue Cutter Salmon P. Chase was used for cadet training from 1878 through 1912.

Upon graduation cadets received commissions as third lieutenants in the Revenue Marine Service. The School of Instruction provided most of the service's new third lieutenants until 1890, when it was forced to close because there were so few openings for beginning officers. The Revenue Cutter Service had no provisions for retiring its older officers, with the predictable result that few retired unless

The Class of 1896, School of Instruction (forerunner of the U.S. Coast Guard Academy).

compelled to do so by serious illness and promotions were held up all down the line. The few openings that occurred each year were filled by surplus Naval Academy graduates until 1894, when the expansion of the Navy absorbed all Academy graduates. In 1895 Congress passed legislation retiring some Revenue Service officers and the ensuing promotions left many openings in the grade of third lieutenant. To fill the vacancies, the School of Instruction resumed operations.

The School of Instruction continued to offer two years of training, but now only candidates who had completed their general scholastic work were admitted. As cadets they concentrated on technical and professional subjects. The *Chase*, lengthened forty feet to make room for a total of twenty-five cadets, returned to service as the school's training ship, moving with the cadets to various southern ports in the winter because the school's former winter quarters at New Bedford were no longer available. In 1900, however, the School of Instruction for the Revenue Cutter Service opened winter quarters at Arundel Cove, Maryland.

To better prepare officers for duty in a Revenue Cutter Service, whose missions were expanding and whose vessels were becoming more complex, Congress, in 1903, extended the training period at the School of Instruction from two to three years, allowing more time for the study of scientific subjects. In 1906 a six-month cadetship (later extended to a full year) was inaugurated for engineers who had previously been commissioned directly from civilian life. They took indoctrinational courses and studied advanced engineering. A new training ship, the *Itasca*, replaced the *Chase* in 1907. The *Itasca*, powered by steam with auxiliary sail, gave the engineering cadets practical experience in ship operation.

Although the Revenue Cutter Service's cadet corps was not large—classes ranged in size from ten to twenty young men—the facilities at Arundel Cove proved inadequate and generally unsuitable for school purposes. Better quarters were located at Fort Trumbull at

Cadets in review at the U.S. Coast Guard Academy, New London, Conn.

New London, Connecticut, and, in 1910, the School of Instruction moved to its new home. In 1914, following improvements in the curriculum, the Revenue Cutter Service changed the name of its school to Revenue Cutter Academy. A year later it became the United States Coast Guard Academy.

During World War I, the Coast Guard Academy, like the other service academies, graduated its classes early, as it was to do again during World War II. The disruption during World War I was short, however, and the cadets soon resumed their three-year program of studies at New London and summer cruises aboard the Academy's training ship. The postwar training ship was the *Alexander Hamilton*, a steam-propelled auxiliary, three-masted barkentine that had begun her career in 1898 as a Navy gunboat. She could accommodate up to a hundred cadets who slept in hammocks.

A training ship as large as the *Hamilton* had become necessary because the Academy was growing, so much so that Fort Trumbull was badly overcrowded. In 1929, after repeated requests from the

Administration Building at Coast Guard Academy is named for Alexander Hamilton, the first Secretary of the Treasury.

Winter snow covers parade ground near Alexander Hamilton Hall at Coast Guard Academy in New London, Conn.

Satterlee Hall at Coast Guard Academy houses classrooms and laboratories where cadets are trained as future coast guard officers.

Coast Guard, Congress appropriated $1,750,000 for "such buildings as [the Secretary of the Treasury] may deem necessary for the purposes of the United States Coast Guard Academy."

In 1932 cadets and instructors moved into new quarters on the west bank of the Thames River in New London. Shortly before the move, the length of the Academy course was extended to four years. Line, or regular, and engineering cadets no longer pursued separate programs; all cadets studied both line and engineering subjects. The curriculum was revised to meet the standards set by the nation's civilian colleges, and in 1940 the Academy was accredited by the Association of American Universities, and authorized to grant the degree of Bachelor of Science.

Today, approximately 1100 cadets are enrolled in the Coast Guard Academy. They come from every state of the union, with a few special cadets coming from the American Republics and the Republic of the Philippines. Appointments at the Academy are made solely on the basis of an annual, nationwide competition; there are no congressional appointments and no geographic quotas.

To qualify for an appointment to the Coast Guard Academy, a young person must be between seventeen and twenty-two years of age, a United States citizen, unmarried, of good character, and must meet certain physical and scholastic requirements. If accepted as a cadet, he or she has a choice of thirteen academic options: general engineering, ocean engineering, electrical engineering, nuclear engineering, marine engineering, civil engineering, mathematics, computer science, chemistry, ocean science, economics/management, physics, and history/government. Athletics and military training are required of all cadets.

During their four years at the Academy, cadets take three cruises of about two weeks each on the training ship *Eagle,* a three-masted auxiliary bark, and two seven-week cruises on Coast Guard cutters. On the cutters the cadets study gunnery, antisubmarine warfare procedures, engineering, and cutter operation.

While at the Academy a cadet receives one-half the base pay of an ensign with less than two years service. The money is used to pay for uniforms, textbooks, and other training expenses.

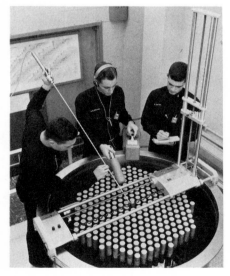

Three first-class cadets conduct experiment with sub-critical nuclear reactor installed in McAllister Hall at U.S. Coast Guard Academy.

Viewed from her port side is the 295-foot, three-masted U.S. Coast Guard Academy training bark Eagle under full sail. This wind-jammer has been the cadets' training ship since 1946.

Climbing the riggings of the training bark Eagle, Coast Guard Academy cadets learn the difficulties of real-life seamanship and sailing.

Seagoing cadets learn firsthand sailing aboard the training ship *Eagle* operated by the Coast Guard Academy.

When graduated, a cadet is awarded a Bachelor of Science degree and a commission as ensign in the United States Coast Guard. The cadet is obliged to serve for five years as a commissioned officer in an organization that is like no other in the range of its functions, combining as it does such varied activities as saving lives, protecting property, enforcing the law, scientific investigation, and preserving national security. The Coast Guard Academy has prepared the cadet to take his or her place as a leader in this unique organization.

The formal statement of the Academy's mission was drafted in 1929 by Superintendent Harry C. Hamlet. It remains in effect today with only an editorial change brought about by the addition of women to the cadet corps in 1976.

To graduate young men and women with sound bodies, stout hearts, and alert minds, with a liking for the sea and its lore, and with that high sense of honor, loyalty, and obedience which goes with trained initiative and leadership; well grounded in seamanship, the sciences, and the amenities and strong in the resolve to be worthy of the traditions of commissioned officers in the United States Coast Guard in the service of their country and humanity.

Cadet on board U.S. Coast Guard training bark Eagle aims a sextant at the sun to determine longitude and latitude during summer cadet practice squadron cruise across Atlantic Ocean.

Cadets practicing at the wheel on board the Eagle at sea.

THE BUILDING OF A LEADER

One of the primary goals of the Academy is to develop in each graduate those leadership qualities and skills so necessary to assume the responsibilities and duties of commissioned Coast Guard officers. The environment at the Academy is more than pure academics and the teaching of professional skills; it is a living laboratory designed to develop in each cadet a sense of honor, integrity, concern for his or her fellow man, and devotion to duty, which has come to be expected of all commissioned officers.

During the first summer and academic year, the cadet will be introduced to the honor concept, to barracks life, and to living and performing in a very regulated, supervised, and strenuous environment. As the cadet advances in rank in the Brigade, opportunities to lead cadets expand. Of course, with this opportunity and authority come added responsibilities. The course entitled "Human Behavior" begins the formal classroom portion of leadership training. This course is taken just prior to involvement in the summer cadre program. During this program, the cadet will be the leader and teacher for the new entering class of cadets, and will have the opportunity to put into practice those skills and concepts presented in class.

Prior to commissioning, the cadet will also take the course "Introduction to Management," which stresses the role of the individual, group, organization, and leadership in the total management process. All cadets also take a course entitled "Legal Systems," which stresses the fundamental nature of law and affords essential professional training in military justice and nonjudical punishment. These two courses round out the formal classroom training to develop necessary leadership skills more fully.

In addition to the formal classroom work, leadership programs are conducted on Saturday mornings by the commissioned battalion officers. During these programs, case studies are discussed with officers who have worked in the Coast Guard and have firsthand knowledge of the challenges and responsibilities that await the cadet upon graduation. The summer training cruises provide extensive practical experience for cadets as they work in roles of petty officers or officers.

During the four years, the cadet will also participate in varsity, junior varsity, and intercompany sports, or other extracurricular programs. These activities offer still another opportunity for displaying and developing sound leadership traits. The entire four years will be spent in a military environment—one of accountability for one's own actions, as well as for the cadets under his command. The total Academy experience is one that instills those traits and skills so necessary to be a leader in today's Coast Guard, and in a rapidly changing technically-oriented society.

In many ways, from appearance and academics to sports and social activities, the Coast Guard Academy resembles a small New England college. But there the resemblance ends. The Academy is a military

Cadet marching off demerits at U.S. Coast Guard Academy.

academy where each and every cadet is subject ot the Uniform Code of Military Justice, as well as an administrative disciplinary system prescribed in *Cadet Regulations.* One soon finds out that the cadet has to live up to the rules and regs.

The nature of any military organization requires that each individual and each unit be responsive to orders from above. This is known as the chain of command, and following the chain of command is absolutely essential to the smooth operational efficiency of all military systems.

All phases of cadet life are governed by the *Cadet Regulations* and the *Procedure and Equipment Manual.* All cadets are required to be familiar with the *Regulations* and *Manual* and to observe them in letter and spirit.

The discipline of the Academy trains one for effective leadership and obedience to authority. The cadet must learn the meaning and spirit of leadership from both the point of view of receiving orders and giving them.

At the top of the chain of command is the Superintendent. Then comes the Commandant of Cadets, whose position is similar to Dean of Students. Under him are Battalion officers, who are Coast Guard officers. They serve as disciplinarians, counselors, and administrative assistants to the Commandant of Cadets.

Underlying the overall philosophy of military discipline is the Honor Concept, a basic, vital force in cadet life. Cadets at the Academy are men and women of honor who neither lie, cheat, steal, nor attempt to deceive.

Color Guard passing in review.

The Corps of Cadets

When one enters the Academy he or she becomes a member of the Corps of Cadets, which provides the opportunity to develop the qualities of leadership essential to a commissioned officer.

The Corps of Cadets is to a large degree self-governing, and the cadets are given maximum responsibility and authority. The Brigade Commander, under the advice of the Commandant of Cadets, directs the daily requirements of inspections, formations, watches, military appearance, spirit, and performance of the Corps of Cadets.

Experience in military command is gained daily by the three upper classes in the exercise of their individual responsibilities within the Brigade. Additional experience in leadership is gained during summer cruises, when upperclassmen serve as officer of the deck, engineering officer of the watch, and master at arms. With minimum supervision from commissioned officers, they are largely responsible for commanding the cruise vessels.

Unlike students at other colleges, a Coast Guard cadet must participate in a program of military training and discipline administered by upperclassmen and officers. Prompt obedience to orders is required and cadets are subject at all times to the military code of justice which is an internal judicial system.

The way the cadet budgets time is extremely important. In addition to maintaining satisfactory grades in all subjects, the cadet must demonstrate by conduct and participation in Cadet Corps activities that he or she has the qualities of character and leadership required of future officers.

Obviously, the Coast Guard Academy is not for everyone. Some highly qualified young men and women might well be advised to go elsewhere for higher education. If they are thinking of trying for appointment, they should be honest with themselves concerning their motives and their abilities to undertake strenuous disciplines. The applicant would be well advised to discuss his or her intentions with parents and high school guidance counselor first.

Fourth Class System

The Academy provides a four-year program of training and education. The fourth class system, which increases the cadet's authority and responsibilities each year, is the basis for a pyramidal chain of command.

The first of these years, fourth class or "swab" year, is probably the most difficult. The cadet must carry a full academic load, undergo intensive military indoctrination, and learn to follow orders from commissioned officers and upperclass cadets. Military life teaches self-control, discipline, and that respect for authority so essential to a successful career as an officer. To make the transition from civilian to military life as effective as possible, one reports to the Academy eight weeks prior to the start of the fall academic semester to learn the proper behavior.

Like at all service academies, women cadets at the Coast Guard Academy are filling more and more important positions in student brigades.

Each year thereafter, the class is given increased authority and command responsibility; eventually, one member of the fourth class, the freshman class, will become Brigade Commander.

Women at the Coast Guard Academy

On June 28, 1976, for the first time in Coast Guard history, women entered the Academy. This dramatic change in the traditions of the Academy was announced by the Commandant of the Coast Guard several months prior to the legislation authorizing women to admission at Annapolis, West Point, and the Air Force Academy.

Beginning with the class of 1980, which entered in 1976, young women became eligible for appointment to the Academy. They undergo basically the same training as their male counterparts, the only differences occurring in the physical education programs. The academic curriculum is the same, as is the leadership and professional training.

Women have been fully integrated into the cadet barracks, but they have women roommates and separate facilities, such as showers, toilets, and sinks. Women's hair length is short, off the collar, to facilitate easy care. Cosmetics and jewelry are limited to the essentials (wrist watches, post earrings for pierced ears, and minimal makeup). Some physical aspects of a cadet's rigorous routine have been slightly modified, but the program is still challenging and demanding. In all other respects, a young woman's role as a cadet is similar to a man's.

Cadet Allowances

A cadet in the Coast Guard receives an allowance provided by law (the allowance is equal to one-half the base pay of an ensign with less than two years of service). At present this equals over $4000 a year, plus a daily food allowance. A cadet's allowance is not a wage or salary; it is money furnished by the government for uniforms, equipment, textbooks, and other expenses incidental to training. A cadet's allowance covers all the cadet's expenses, and is disbursed and expended only as directed by the Superintendent. Any funds remaining in a cadet's account are given to the cadet upon graduation.

A cadet receives the following amounts each month in a checking account for his personal use: first semester, $50; second semester, $75; second year, $90; third year, $135; fourth year, $150.

A Day in the Life of a Cadet

It's a full life.

A typical day begins at the crack of dawn, 6:10 A.M., with reveille. Cadets prepare for daily personnel inspection and breakfast; daily inspections assure that rooms are clean and orderly.

Corps of Cadets dine in wardroom of the Chase Hall extention. The windows of the wardroom look upon the Thames River at New London, Conn.

Memorial Chapel is where cadets attend religious services at Coast Guard Academy.

Morning classes begin at 7:50 and end at 11:40. Lunch is at 12:00, followed by afternoon classes from 12:50 to 3:30. The time between 3:30 and 6:00 P.M. is spent in intramural or varsity sports, extracurricular activities, and extra classes of academic instruction for those cadets who need additional help.

Dinner is at 6:30. The hour between 7:00 and 8:00 is set aside for personal use, during which cadets may study or attend to personal matters. Study time is set between 8:00 and 9:45. Taps is at 10:00 but cadets may continue to study until midnight.

Saturday mornings are spent in professional training and indoctrination. During the fall and spring months, formal regimental parades and inspections are held on Friday afternoons.

Liberty, which means permission to leave the Academy grounds, is granted on Saturday afternoon and evening, and again on Sunday. First classmen are given liberty on Wednesday afternoons at 4:00, and all upperclassmen get liberty on Friday at 4:00 P.M.

Provision is made to attend religious services of one's faith. Catholic and Protestant chaplains provide the opportunity for Christian worship and study. Jewish cadets and those from other religious groups are assisted in locating facilities for worship in the civilian community. Attendance at religious services is voluntary.

All cadets live on campus at the Academy. Each cadet will share a room with another cadet and both are responsible for its cleanliness and neatness.

Not every day in the life of a cadet is spent at the Academy. During the year a cadet is given ample opportunity for leave and holidays.

Professional and Military Training

Professional and military training does not end with the academic courses and Saturday morning orientation period. Between Commencement Day and Labor Day, with time out for a few week's

Chapel window is dedicated to "Freedom of Speech" at Coast Guard Academy.

leave, the cadets spend summers putting what has been learned during the academic year to practical use. They take training cruises to many of the historic, exotic ports of the world, a fabulous experience not shared by many young people in their college years.

The cadets receive professional training at various Service Training Centers, and work with operating units to enhance their professional competence.

One of the training ships making summer practice cruises is the three-masted auxiliary bark, *Eagle*. The *Eagle* has a diesel engine, electricity, evaporators, the latest electronic devices for navigation and operations, and all the conveniences of modern ships.

In addition, two or more of the latest Coast Guard cutters are diverted from their regular duty to provide Academy cadets with training in gunnery, antisubmarine warfare procedures, engineering, and the routing of actual operating units.

On these cruises, the cadets obtain a thorough foundation in practical seamanship, ordnance and gunnery, small arms, and communications. The cadets participate in battle problems and search and rescue drills, and learn to control and fire all the primary weapons found on Coast Guard cutters. They apply their classroom knowledge of seamanship, handle small boats, and observe the handling of larger vessels. Upperclassmen navigate ships and command the watch, applying the theories of nautical science.

During the first summer as a fourth classman, known as swab summer, the emphasis is on indoctrination. In a sense, it's similar to "boot camp." The cadets take courses in all phases of military activity. They are required to pass a physical fitness and swimming test. Cadets not in shape when they arrive for "swab summer," will soon be. Emphasis is placed on physical fitness and competitive sports. The highlight of this first summer is a one-week cruise aboard the *Eagle*.

For the third classman, the second summer will be a long cruise aboard cutters or the *Eagle* (about ten weeks), which may take the cadets to such faraway places as Japan or England.

During the third summer, the cadets are trained in seamanship, navigation, damage control, and fire-fighting procedures. They also may help in training the incoming fourth class swabs, or participate in Boys State and Girls State programs throughout the country.

Two weeks of Coast Guard aviation orientation are highlights of the second class summer program. Cadets study the basic theory of flight and operational employment of aircraft. They have an opportunity to handle the controls, navigate aircraft, and experience firsthand the problems faced by pilots.

During the first class summer program, the cadet's senior year summer, they take a five-week cruise, frequently to foreign ports. They also spend five weeks at various Coast Guard installations ashore, working in the field on actual Coast Guard operations.

Athletic Program

Athletics are compulsory at the Academy and are as much a part of

Varsity intercollegiate football ranks among favorite participant and spectator sports at the U.S. Coast Guard Academy.

A complete intramural sports program for cadets is featured at the New London, Conn. campus of the U.S. Coast Guard Academy.

the educational program as academics and military indoctrination. Physical fitness, agility, stamina, and a competitive spirit are important. In addition, athletics provides recreation, develops teamwork, and becomes the basis for a lifelong physical fitness program.

All cadets are required to participate in a physical education program that includes the following:

first year: foundations of physical activity, survival swimming, gymnastics, wrestling or dance;
second year: tennis, handball or synchronized swimming, volleyball, advanced swimming;
third year: personal defense, golf, life saving, badminton;
fourth year: SCUBA, water safety instruction, ecology sports, skiing, golf, elected sports, leadership.

Special remedial swimming and exercise programs are conducted for cadets who score low on entrance tests of swimming and physical fitness.

In addition to the required physical education classes, cadets must participate in either intramural or intercollegiate athletics. The Academy, as a member of the NCAA and the New England Intercollegiate Athletic Association, competes with other member colleges across the country.

Football is one of the popular sports, and on many fall afternoons, the cheers of the Cadet Corps resound through colorful Cadet Memorial Field. Other popular sports on the fall schedule are sailing, soccer, cross-counrty, and tennis.

The Academy is a member of the Intercollegiate Yacht Racing Association, and sails against national and international competition. The Sailing Squadron regularly races a fleet of sloops, dinghies, and

Cadet band at the Coast Guard Academy plays at sports and special events throughout academic year.

119

Wrestling is also a very popular sport with the cadets.

Pleasant living quarters make off-duty a pleasure for cadets at the Coast Guard Academy.

Coast Guard cadets get experience in sailing light craft—as well as large—during their four years training at the Academy.

oceangoing yachts both socially and competitively.

During the winter months, the Academy's new field house, Roland Hall, one of the finest facilities of its kind in the country, is the scene of exciting competition in basketball, wrestling, swimming, indoor track, gymnastics, and rifle and pistol matches. In the spring, sailing, track, golf, crew, and baseball provide plenty of activity for cadets.

An extensive program of intramural sports is carried out during recreation hours.

Extracurricular Activities

Formal and informal dances, parties, movies, and other social functions are scheduled regularly. Cadets find an outlet for their musical talents in a strong program of musical activities, which includes: the Cadet Protestant and Catholic Choirs and Glee Club; the Coast Guard Academy Idlers, a singing group that has appeared on national television; the Windjammers Marching Band; the Cadet Dance Orchestra; and several small groups playing contemporary music. A cadet who plays a musical instrument should bring it to the Academy.

The Karate Club, Bowling Club, Debate Team, and Radio Club, which operates WICGA, Bicycle Club, Hockey Club, Rugby Team, and the Christian Fellowship are typical of over twenty extracurricular organizations active at the Academy. *Howling Gale,* the Cadet magazine, and *Tide Rips,* the Academy yearbook, offer experience and pleasure to cadets with interests in writing.

Academic Departments

The Academic Division consists of eight departments under the direction and supervision of the Dean of Academics: Applied sci-

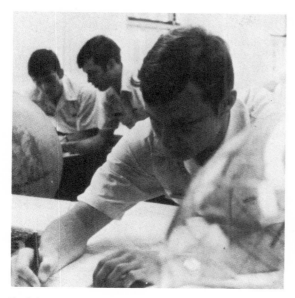

Studying U.S. ocean geography is important for Coast Guard Academy cadets.

Academic courses are both difficult and rewarding at the Coast Guard Academy.

Sea navigation is featured in the academic program for cadets at the U.S. Coast Guard Academy.

ences and engineering, nautical science and law, mathematics, physical and ocean science, economics and management, computer science, humanities, and physical education. These departments are staffed by officers of the U.S. Coast Guard, and a body of permanent professors, both civilian and military.

The courses presented by these experienced educators assure Academy graduates of a broad foundation in the liberal arts and social sciences, in physical science, in engineering, and in those professional studies that specifically prepare graduates for their careers in the Coast Guard.

Major Courses of Instruction

The United States Coast Guard Academy recognizes that the majority of Academy graduates must have an engineering or scientific background to qualify them to meet the needs of the Coast Guard.

The Academy academic program, responsive to both the needs of the Coast Guard and the interests of the Corps of Cadets, currently consists of the following majors:

marine engineering,
ocean engineering,
electrical engineering,
civil engineering,
marine science,
mathematical sciences,
physical sciences,
management, and
government.

During the second semester of his or her first year, each cadet selects a major and is assigned to an academic advisor. Enrollment in a specific major is sometimes limited. When such a condition arises, selection of those students desiring admittance into a major is made by the Dean of Academics, and is based solely on the students' prior academic performance at the Academy. Throughout the four-year academic program all majors require that a cadet pass a core program of twenty-seven courses in addition to those required by the major.

As cadets progress in their four-year sequences, they may select additional electives that allow them to pursue even further individual academic interests.

Each major provides a sound undergraduate education in a field of interest to the Coast Guard, and prepares the cadet to assume initial duties as a junior officer. Upon graduation, the cadet is awarded the degree of Bachelor of Science, and, if physically qualified, is commissioned by the President of the United States as an ensign in the Coast Guard. Applicants are cautioned, however, that the opportunity for specific postgraduate work and service assignments are dictated by the needs of the Coast Guard.

Validation

Validation is the process through which applicants can receive credit for fulfilling the requirements of specific courses in the Academy curriculum before entering the Academy. Each course validated permits a cadet to pursue an advance course in the same subject or an extra elective. If the applicants have performed satisfactorily on College Entrance Examination Board Advanced Placement Tests and/or have satisfactorily completed courses at other institutions of higher learning; or advanced or college-level courses in high school, they are eligible for validation. Validation is based on a local screening process.

Through the Academy's validation program the cadets may tailor their academic programs to a personal level of achievement. This gives the cadets the opportunity to enrich their undergraduate education through the proper use of electives made available by validation.

Honors Placement

The honors placement program permits one to reach the highest level of academic achievement by enrolling in the honors (advanced) sections of certain courses. Honors programs are designed and administered by the individual departments.

What Classes Are Like

The methods of instruction at the Academy are quite varied. Occasionally the members of an entire class will meet together for general lectures. Usually, classes for required courses consist of eighteen to twenty-two cadets; those for elective courses as few as six. In order to supplement its classroom instruction, the Academy faculty makes considerable use of pertinent audiovisual aids, laboratory exercises, the computer center, and library resources.

Individual Assistance and Counseling

The Academy has a program of individual assistance and counseling to help each cadet achieve academic success. Additional instruction is available whenever required. Academic assistance and guidance on an individual basis are provided by company officers and instructors. Counseling and personal assistance are also available to help cadets through normal developmental and emotional concerns. This is provided by a civilian counselor and the Academy's two chaplains.

Grades and Honors Lists

The assigned grades for each course are based on the cadet's daily work over the sixteen-week semester and the individual's score on the final examination.

At midsemester and at the end of each term a quality point average is determined with A = 4, B+ = 3.3, B = 3, C+ = 2.3, C = 2, D = 1, and F = 0. A cadet must obtain at least a 2.0 cumulative quality average to graduate.

Cadets are encouraged to strive for academic excellence by the prospect of attaining the Honors List, and of receiving the additional privileges granted in conjunction with that achievement. Honors are awarded to cadets who achieve not less than a 3.15 quality point average, with no mark in any three-hour course less than 2.0 (C).

Academy Scholars Program

The Academy scholars program offers a special intellectual challenge to cadets in their fourth year who have demonstrated outstanding scholastic abilities. Those selected are given special recognition and academic privileges, which enable them to pursue, under faculty guidance, individually selected projects and special research. Recent projects were "Analog Computer Solution to the Dynamically Developed Force at the Bow of an Icebreaker," "An Analysis of Numerical Methods Employed in Naval Architectue," and "Ionic Conduction Communications."

Coast Guard Duty

First of all, the United States Coast Guard is a military service. It serves the country's needs first and last. In time of peace, the Coast Guard functions as a humanitarian arm of the Department of Transportation. In time of war, it serves as a highly-specialized part of the United States Navy.

Former President Lyndon B. Johnson presents graduation diploma and ensign's commission to young cadet upon graduation at U.S. Coast Guard Academy.

Any officer in a military service must be well trained and professionally qualified for leadership, but this is especially true in the Coast Guard, because it is the smallest of the military services. It is divided into small units of professionals, sometimes so small that command responsibility is often thrust upon officers early in their careers. Producing officers who can meet this challenge is the function of the United States Coast Guard Academy. The Academy's sole purpose is to provide career officers for the United States Coast Guard.

The Coast Guard Academy graduate must serve a minimum of five years as a Coast Guard officer. Four years at the Academy and subsequent service as a Coast Guard officer provide many satisfactions to young people who enjoy variety and excitement, who are interested in technical and administrative work, and who have an essentially idealistic and humanitarian outlook on life.

The first experience as a newly commissioned Coast Guard officer will probably be at sea. The new Coast Guard officer may be assigned to one of the larger cutters for fisheries patrol and search and rescue, or to one of the modern icebreakers involved in the Arctic and Antarctic, or perhaps to one of the large buoy tenders that service aids to navigation along the coasts and on the Great Lakes.

No matter where, the new ensign will find that a Coast Guard officer is given greater responsibility at a much younger age than most civilian contemporaries.

Postgraduate Opportunities

Coast Guard officers are currently enrolled in the following advanced and specialized courses (availability varies with Coast Guard needs):

advanced electronics
aerospace mechanical engineering*
aircraft maintenance
Air War College
Armed Forces Staff College
aviation electronics engineering*
aviation engineering administration*
chemical engineering*
civil engineering
communications engineering*
electronic engineering*
engineering physics*
environmental management*
financial management*
humanities*
Industrial College of the Armed Forces
law*
management*
management (data processing)*
management and industrial engineering*
management (personnel administration)*

mathematics*
Merchant Marine industry training
National War College
naval engineering*
Naval War College
nuclear engineering*
oceanography*
operation analysis*
public administration*
science (chemistry or physics)*
systems analysis*

After the initial tour of sea duty, a Coast Guard officer is encouraged to apply for a postgraduate education or specialized training in the field of his choice.

The officer may apply for aviation training and, if selected, will be assigned to flight training under the direction of Coast Guard and Navy instructors. After earning his wings, he will then be assigned to one of the Coast Guard Air Stations. The particular graduate specialties available for choice are dictated by the needs of the Coast Guard.

The Coast Guard, in conducting its postgraduate education program, enrolls officers in some of the leading colleges and universities throughout the country. Sixty percent of all Academy graduates receive further formal education during their first five years of service. Officers are selected for postgraduate education and specialized training on the basis of scholastic standing at the Academy, performance as an officer, and professional qualifications.

Most graduates elect to continue their careers as Coast Guard officers after they complete the required five years of service. They find a Coast Guard career worthy of their abilities, for the Coast Guard, unlike other services, has a vital job to perform in peacetime as well as in war.

There is a great deal of stimulus in the opportunity to travel, the variety of assignments, opportunities for specialization in different, exacting fields, plus the security and fringe benefits of an officer's life. But service as a Coast Guard officer also means long periods at sea with seven-day work weeks and long, irregular hours. It means responsiblity for the safety of a ship and its crew. In certain moments, it can mean facing actual physical danger. There is no question but that it is demanding. Nevertheless, to the well-motivated, well-prepared young officer who thrives on the age-old challenge of men and ships against the sea, it is exciting and meaningful.

While advancing in rank and experience, the Coast Guard officer will be assigned to an administrative billet ashore. In this capacity, the officer begins to apply his or her education and experience to solving problems concerning water pollution, boating safety, and law enforcement.

* Leads to an advanced degree.

126

How To Apply For Appointment

The U.S. Coast Guard Academy differs from the other service academies, in that appointments are tendered solely on the basis of an annual nationwide competition. There are no congressional appointments, state quotas, or special categories. All applicants, whether civilians or members of the armed forces, participate in the competition on an equal basis.

The competition for appointment as a cadet is based on high school rank, performance on either the College Board Scholastic Aptitude Test (SAT) or the American College Testing Assessment (ACT), and leadership potential as demonstrated by participation in high school extracurricular activities, community affairs, or part-time employment.

To be considered for appointment, applicants must take either the College Board Scholastic Aptitude Test (SAT) not later than December; or the American College Testing Assessment (ACT) not later than November.

An important reminder: an applicant must submit a formal application form postmarked no later than December 15.

Visiting the Academy

The Coast Guard Academy is open to visitors daily from 9:00 A.M. to sunset. If a prospective candidate desires an interview with a member of the admissions staff, the admissions office is open Monday through Friday from 8:00 A.M. to 4:30 P.M., excluding national holidays.

On Friday afternoons at 1:00, there is a special admissions presentation, including a slide show, question-and-answer period, and tour of the campus. Appointments for an interview or the Friday afternoon presentation may be arranged by writing to the Director of Admissions, U.S. Coast Guard Academy, New London, Connecticut 06320, or by calling (203) 443-8463, extension 603.

THE ACADEMY PHILOSOPHY

The Academy, like other colleges, strives to make each graduate a well-informed, contributing member of society, fully prepared to accept life's responsibilities.

But that's just the beginning. Each Academy graduate must also be a qualified officer, capable of leadership. For this reason, the program at the Academy is more rigorous and more demanding than nonmilitary colleges in order to meet the professional needs of the U.S. Coast Guard.

The educational philosophy of the Academy has three basic objectives:

1. to provide by precept and example an environment that encourages a high sense of honor, loyalty, and obedience;

2. to provide a sound undergraduate education in a field of interest to the Coast Guard; and

3. to provide training that enables graduates to assume their immediate duties as junior officers aboard ship.

With these three principles serving as guidelines, the Academy staff has evolved the present curriculum, which is fully accredited by the New England Association of Colleges and Seconday Schools, and which leads to a Bachelor of Science degree and a commission as ensign in the United States Coast Guard.

Questions and Answers About the Coast Guard Academy

Why choose the Coast Guard Academy?

To get a good education, first of all, one that will serve as a foundation for a rewarding career. Each year there will be about 350 young men and women who will win admission to the Acadmey in New London, Connecticut. They will choose from nine majors that can lead to a Bachelor of Science degree. Specifically, these are: marine engineering, ocean engineering, electrical engineering, civil engineering, marine science, mathematical science, physical sciences, management, and government.

The curriculum is accredited by the New England Association of Colleges and Secondary Schools. Accreditation by the Engineer's Council for Professional Development is given marine engineering, ocean engineering, and electrical engineering majors. The curriculum is constantly being updated to keep abreast of modern trends in education. In addition, the Coast Guard offers postgraduate opportunities in many fields related to those above, a good number of which can lead to advanced degrees.

What's the mission of the Academy?

To graduate career officers for the U.S. Coast Guard. That's it in a nutshell, but it implies much more. A Coast Guard officer must be as well trained and professionally qualified for leadership as any military officer, but, because the Coast Guard is the smallest of the military services (only 44,000 strong), its officers often have command responsibility thrust upon them much earlier than would be the case with the other services.

The job of the Coast Guard is saving lives and property. In time of peace, the Coast Guard serves as the humanitarian arm of the Department of Transportation. In a recent year, for instance, it answered approximately 79,000 calls for help, prevented about 4200 deaths, and assisted more than 125,000 people. It also saved property valued at about $300 million. In time of war, the Coast Guard serves as a highly-specialized wing of the United States Navy.

What is life like at the Academy?

In its rustic New England setting, the Coast Guard Academy looks like many small college campuses. Woods, soft rolling hills, and a quiet river provide an exquisite backdrop for college life.

The similarity stops there. Life at the Academy is highly disciplined. All cadets are subject to the Uniform Code of Military Justice, and they all live under an administrative disciplinary system prescribed in cadet regulations.

What is the fourth class system?

This is a program for training and indoctrinating new cadets. It means that they must take orders from upperclassmen as well as from the commissioned officers at the Academy. It sets forth a number of rules for the behavior of fourth classmen, of "swabs," to help make the transition from civilian to military life. This time-honored system teaches self-control, discipline, and respect for authority. Each class is given increased authority each year, until one day a former member of the fourth class will be Brigade Commander.

What's a typical day?

All cadets live on campus at the Academy. Each cadet shares a room that is neatly furnished with everything required for comfortable living and studying. There are daily inspections to make sure the rooms are clean and orderly. Cadets follow a carefully planned routine of activities.

A typical day begins at 6:10 A.M. when reveille is sounded. The cadet is awakened and prepares for daily personnel inspection and breakfast. Morning classes start at 7:50 and go until 11:40. Lunch is at 12:00 followed by afternoon classes from 12:50 until 3:30. The hours between 3:30 and 6:00 P.M. are given over to intramural or varsity sports, extracurricular activities, or extra classes if cadets need individual help. Dinner is at 6:30 and there is an hour set aside from 7:00 to 8:00 for free activities. Study time is set from 8:00 to 9:45. Taps marks the end of the day at 10:00, but further study is permitted until midnight.

It's a full life, as you can see, and it's one reason why Coast Guard officers are tough, disciplined, exacting leaders.

On Friday afternoons in the fall and spring, there are formal regimental parades. Saturday mornings are devoted to professional training. Liberty, which means permission to leave the Academy grounds, is granted on Saturday afternoons and evenings, and again on Sundays. Senior upperclassmen are also given liberty on Wednesday, and all upperclassmen on Friday afternoons at 4:00.P.M.

What about athletics and extracurricular activities?

Athletics is as much a part of the Academy program as academics and military indoctrination. Physical fitness is important to an officer, and all cadets are required to participate in a physical fitness program designed to teach teamwork and provide a lifelong basis of good health and physical well-being. All cadets must pass certain minimum physical aptitude standards and swimming tests, and take part in either intramural athletics or varsity sports in the P.M.

And the extracurriculars?

Dances, music, theater, movies, sailing, photography, art, and writing are a few of the outside activities scheduled regularly at the Academy. Whatever your personal interest, you can find an outlet for it. The academy has several nationally known musical groups, as well as clubs for a variety of such other activities as bowling, karate, debating, theater, radio commentating, jazz, and more.

Are cadets paid at the Academy?

A cadet in the Coast Guard receives an allowance provided by law (equal to one-half the base pay of an ensign with less than two years service). Right now this equals over $4000. It is money furnished by the government to pay for uniforms, equipment, textbooks, and other incidental expenses. It is not unusual for cadets to have a considerable sum of money saved up upon graduation.

5

The United States Merchant Marine Academy

The purpose of the United States Merchant Marine Academy is to train, develop, and impart to outstanding young men and women the academic background, training, motivation, and ambition to serve as officers in the American merchant marine and as leaders in the U.S. maritime industry. Graduates of the Academy are making significant contributions to the maritime industry as merchant officers, shipping company executives, naval architects, admiralty lawyers, marine underwriters, oceanographers, and career officers in the U.S. Navy and Coast Guard.

The Maritime Administration of the U.S. Department of Commerce operates the U.S. Merchant Marine Academy at Kings Point, N.Y., where young people are trained to become merchant marine

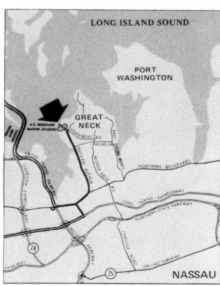

Maps of U.S. Merchant Marine Academy.

officers. It also administers a grant-in-aid program for those maritime academies operated by the states of California, Maine, Massachusetts, Michigan, New York, and Texas.

Two-thirds of the earth's surface is covered by water and, since the earliest days of recorded history, the development of ocean transportation and commerce has coincided with the development of mankind. Today, as a result of advanced technology in ship design, modern management techniques, reliance of all nations on balanced international trade, and growing interest in the oceans as a source of natural resources and food, the challenge of the sea is greater than ever.

An adequately sized U.S. merchant fleet of highly productive ships guarantees our nation access to foreign sources of raw materials and to foreign markets for sales of our manufactured goods. In addition, such a fleet is vital for our national security for, in time of war or national emergency, our merchant ships are our "fourth arm of defense," and carry the brunt of the task of delivering military supplies across the oceans to our forces and allies overseas.

The major component of our nation's maritime industry is the U.S. merchant marine—the fleet of privately owned, American-flag vessels that transports goods, mail, and passengers to all ports of the world. Yet a strong maritime industry consists of more than oceangoing ships. It also includes large numbers of tugs, barges and river craft, steamship companies, ports and terminals, shipyards, marine insurance underwriters, ship chartering firms, admiralty lawyers, and a vast array of other specialized firms directly or indirectly engaged in maritime-related commerce, engineering, research, and management endeavors.

More than anything else, a productive merchant fleet and a strong maritime industry require people—men and women who are intelligent, ambitious, well educated, and competent. The prime purpose of the Merchant Marine Academy is to assure that such persons are available to meet the challenges of the present and the future.

HISTORY OF THE ACADEMY

The United States Merchant Marine Cadet Corps was established on March 15, 1938, following passage of the Merchant Marine Act of 1936. Training was first given aboard merchant ships and later at temporary shore establishments pending the acquisition of permanent facilities. The Walter P. Chrysler estate at Kings Point, New York, was selected as the permanent site for the Academy in March 1942, and construction was begun the following May. Fifteen months later the task was virtually completed, and the U.S. Merchant Marine Academy was dedicated on September 30, 1943.

World War II required the Academy to forego normal operation and devote all its resources toward meeting the emergency personnel needs of the merchant marine. The enrollment was increased to 2700, and the planned course of instruction was reduced from four years to two. Notwithstanding the war, shipboard training continued to be an

Wiley Hall—The former Walter Chrysler Residence.

The Merchant Marine Academy at Kings Point, New York, was founded September 30, 1943.

integral part of the Academy curriculum, and midshipmen served at sea in combat zones the world over. Two hundred twelve midshipmen and graduates gave their lives in service to their country and many others survived torpedoings and bombings. Seven midshipmen and one graduate were awarded the Merchant Marine Distinguished Service Medal, the nation's highest decoration for conspicuous gallantry and devotion to duty, and, by the end of the war, the Academy had graduated 6634 officers.

In the closing days of World War II, plans were formulated to establish a four-year, college-level program to meet the peacetime needs of the merchant marine. At the end of the war, in August 1945, the four-year course was immediately instituted with the September class of midshipmen.

The Academy has since grown in stature and has become one of the world's foremost institutions in the field of maritime education. Authorization for awarding graduates the degree of Bachelor of Science was granted by the Eighty-first Congress in a law approved August 18, 1949, and the Academy was fully accredited as a degree-granting institution by the Middle States Association of Colleges and Secondary Schools on November 26, 1949. The Academy was made a permanent institution by an Act of Congress on February 20, 1956, and its operation was placed under the authority of the Department of Commerce. Today, Kings Point graduates are serving with distinction in all sectors of the maritime industry—as ships' officers, steamship company executives, admiralty lawyers, marine underwriters, naval architects, oceanographers, and career officers in the U.S. Navy and Coast Guard. With the 1974–75 academic year, the Academy opened its doors to young women, who will find broad opportunities for satisfying careers in the maritime industry.

WOMEN AT USMMA

When the Merchant Marine Academy accepted women as midshipmen in 1974, it was the first of the five federal academies to become coeducational. The standards for accepting women are high. Women must meet the same selection criteria as men, and they must also meet the commissioning standards of the U.S. Naval Reserve. At the

133

When the Merchant Marine Academy accepted women as midshipmen in 1974, it was the first of the five federal academies to become coeducational.

The maritime industry is a growing field for women, and one in which they can find significant careers.

At the Merchant Marine Academy, men and women take the same courses, attend the same classes, and participate in the same activities.

Academy, men and women take the same courses, attend the same classes, and participate in the same activities. This academic and military training is demanding, and women must be in excellent physical condition. All varsity sports are open to women, with the exception of football and wrestling. Club sports for women are fencing, gymnastics, and the martial arts. There are also many extracurricular activities, social mixers, and dances. Under the Shipboard Training Program, women are assigned to merchant ships during one-half of each of their sophomore and junior years. The Academy expects that, upon graduation, men and women will have identical career opportunities open to them. Many outstanding young women are being attracted to the Academy today. As one woman student expressed it, "I wanted a college that would give me not only a good education but also a good chance of finding a job." On board ship or ashore, the maritime industry is a new and growing field for women, and one in which they will find significant careers.

LOCATION

The Academy is located on sixty-eight acres of land at Kings Point, on the north shore of Long Island Sound, approximately twenty miles east of New York City.

BUILDINGS AND FACILITIES

The Academy campus and facilities were carefully planned for a normal enrollment of approximately 1000 midshipmen (the term is

The Merchant Marine Academy today ranks as one of the world's foremost institutions in the field of maritime education.

The War Memorial.

now applied to women as well as men). The design of the buildings is simple yet functional, and the campus has been laid out to take full advantage of the picturesque landscape of the north shore. The buildings and walks are named after men whose deeds brought fame to the merchant marine or to the Academy.

On the slope looking toward Long Island Sound stands a monument erected to the memory of the two hundred twelve midshipmen and graduates who lost their lives at sea during World War II. The War Memorial serves as the focal point for the western part of the campus.

Around the monument, adjoining the Sound, are grouped the following buildings and facilities: indoor and outdoor swimming pools; boat basin and pier facilities; the Wooster Building, housing the Public Works Department; the Fitch Building, housing the Finance and Supply Department; and Samuels Hall, with classrooms and laboratories for teaching nautical science. A beautiful Memorial Chapel, honoring all men of the merchant marine, stands on a grassy knoll to the south of the War Memorial.

Wiley Hall, facing Long Island Sound, is the center of administrative activities, and contains the offices of the Superintendent, Assistant Superintendent, Academic Dean, Commandant of Midshipmen, and other members of the administrative staff. East of Wiley Hall lies the center of the campus, marked by one of the nation's tallest flagpoles, 176 feet 6 inches high. Surrounding the flagpole are Fulton Hall, the engineering and science building; Bowditch Hall, housing the Department of Humanities, the Department of Law and Eco-

Library facilities at the Merchant Marine Academy are excellent.

nomics, and the auditorium; and the Schuyler Otis Bland Memorial Library. Delano Hall, the midshipman dining room, and six dormitory buildings—Jones, Barry, Rogers, Cleveland, Murphy, and Palmer halls—complete the circle of buildings enclosing the main campus. The dormitories and dining room are connected by an underground promenade that contains the midshipman lounge and canteen, the uniform and varsity shop, laundry facilities, bank, barber shop, and ship's service store.

On the perimeter of the Academy are the athletic fields; Furuseth Hall, containing the nuclear engineering facilities and the National Maritime Research Center; O'Hara Hall, which has a spacious gymnasium and indoor athletic facilities, as well as naval science classrooms; the Infirmary; and Land Hall, the midshipman recreation and activities building. Other buildings located on the grounds are used as residences by the Superintendent and officers of the administrative staff.

MIDSHIPMAN LIFE

The Regimental Program

Military life at the United States Merchant Marine Academy is a vital part of a midshipman's total educational experience, and all midshipmen are required to meet high standards of conduct and discipline. The regimental program is carefully designed to provide the midshipmen with leadership training and experience, and to develop in them qualities of self-discipline and responsibility for effective citizenship and successful careers as officers and leaders in the maritime industry.

The Class System

Fundamental to the regimental program is the class system of responsibilities, duties, and privileges. First classmen (seniors),

The Merchant Marine military program is carefully designed to develop leadership ability, self-discipline, and a sense of responsibility.

Practising military drills and ceremonies is a part of student life at the Merchant Marine Academy.

under the supervision of the Commandant of Midshipmen and his staff, exercise military command of the Regiment of Midshipmen. Each first classman has three opportunities during his or her senior year to serve in one of the one hundred forty-four Midshipman Officer billets.

The Regiment of Midshipmen, under the overall command of the Midshipman Regimental Commander, is divided into three battalions, each under the command of a Midshipman Battalion Commander.

The student body at the Merchant Marine Academy is organized along military lines as a regiment, under the command of the senior class.

These top-ranking Midshipman Officers work closely with the Commandant in formulating and carrying out policies and procedures relating to all facets of midshipman life, thereby receiving practical leadership experience that helps to develop self-confidence, improves their understanding of human relations, and instills in them a sense of responsibility. First classmen are granted privileges commensurate with their seniority and responsibility.

The juniors and sophomores, called second classmen and third classmen at the Academy, are primarily responsible for assisting the first class in the indoctrination and orientation of the freshmen, who are more commonly known as plebes. The upperclassmen assure that the plebes display proper military bearing and practice correct military etiquette, and indoctrinate them in the history and traditions of the Academy. The privileges granted the second and third classes are less than those enjoyed by the first class but more liberal than those granted to the plebes.

The Honor Concept

The Honor Concept belongs to the Regiment of Midshipmen. It is a midshipman concept designed for the benefit of all midshipmen, and it creates an atmosphere that improves the quality of life and the educational experience of the Academy.

The Honor Concept is so simple that it is contained in one sentence: "A midshipman will not lie, cheat, or steal." This one sentence must be completely accepted and backed by every member of the Regiment, and every member of the Regiment must accept the consequences of any violation of the Honor Concept.

It is of vital importance that all midshipmen understand that the Honor Concept is designed to protect them in their daily living, to give greater value to their degrees, and to instill in them the principles of honesty and integrity that are so essential to a full and rewarding life. The spirit of integrity developed at the Academy will remain as a personal asset to each individual throughout his or her life.

The newly appointed plebes arrive.

138

Plebe Orientation Program

The newly appointed plebe reports to the Academy during the third week in July for two weeks of orientation prior to the beginning of classes in August. During the orientation program, and during most of the remainder of plebe year, the new midshipman undergoes an intensive Program of military training and indoctrination. The life of a plebe is rigorous and demanding—under the class system he or she has the greatest number of obligations and the fewest privileges—but the new midshipman soon becomes well versed in Academy traditions, develops a keen sense of pride and *esprit de corps,* and completes the adjustment to the requirements of the military program.

Acceptance Day Ceremony. Acceptance Day marks the entry of a new class into the regiment of midshipmen.

A Typical Day During Indoctrination

5:40 A.M.	Reveille—all hands
5:45 A.M.	Morning calisthenics—Barney Square
6:10 A.M.	Showers; clean rooms; cleaning stations
6:45 A.M.	Morning mess
7:15 A.M.	All hands clear Mess Hall; cleaning stations
7:35 A.M.	Morning inspection
7:50 A.M.	Call to Colors. Sections muster on Barney Square
8:00 A.M.	Morning Colors
8:10 A.M.	First instruction period begins
9:00 A.M.	First instruction period ends
9:10 A.M.	Second instruction period begins
10:00 A.M.	Second instruction period ends
10:10 A.M.	Third instruction period begins
11:00 A.M.	Third instruction period ends
11:10 A.M.	Fourth instruction period begins
Noon	Fourth instruction period ends
12:15 P.M.	Noon mess formation
1:10 P.M.	Fifth instruction period begins
2:00 P.M.	Fifth instruction period ends
2:10 P.M.	Sixth instruction period begins
3:00 P.M.	Sixth instruction period ends
3:10 P.M.	Seventh instruction period begins
4:00 P.M.	Seventh instruction period ends
4:00–6:00 P.M.	Intramurals; free time
6:00 P.M.	Showers
6:30 P.M.	Evening mess
7:00–9:00 P.M.	Evening activities
9:00 P.M.	Showers
9:30 P.M.	Tattoo; accountability check
9:40 P.M.	Taps
9:45 P.M.	Lights out

From the above schedule, one can readily understand why it is imperative that all appointees report to the Academy in good physical condition.

The Daily Schedule

A midshipman's daily routine begins with reveille, 6:10 A.M. for plebes and 7:10 A.M. for upperclassmen. Breakfast, optional for upperclassmen, is at 6:40 A.M. Following colors formation at 8:00 A.M., midshipmen proceed to class, and classes continue until noon. Lunch is at 12:20 P.M. and afternoon classes are scheduled from 1:10 P.M. until 4 P.M. Since some class periods are assigned as study hours, a midshipman spends about five hours in class each weekday.

The time between 4 P.M. and 6:30 P.M. is a midshipman's free time and is normally devoted to varsity athletics, intramurals, club meetings, extra study, or some form of extracurricular recreation. Following the evening meal an evening study period extends until the evening accountability check at 9:30 P.M. Midshipmen may continue to study in their rooms after 9:30; taps is sounded at 10 P.M. for those midshipmen who wish to retire. Saturday mornings are used for Regimental parades and inspections, but the remainder of each weekend is used for liberty and recreation.

Though discipline is important at the Merchant Marine Academy, midshipmen are granted liberty on weekends and granted leave during Thanksgiving, Christmas, Easter, and the month of July.

Leave and Liberty

All midshipmen are granted approximately two weeks leave during the Christmas holiday and four days leave during Thanksgiving and Easter. In addition, they receive annual leave during the month of July.

First, second, and third classmen are normally granted weekend liberty from 12:30 P.M. on Saturday until 9 P.M. or 10 P.M. on Sunday evening. At least once each quarter, and occasionally more frequently if circumstances permit, midshipmen are granted long weekends commencing Friday afternoon after the last scheduled class.

Plebes are not usually granted overnight liberty. Beginning in September, after an extended period of indoctrination, they are granted liberty from 8 A.M. to 8 P.M. on Sunday. Saturday liberty, granted to approximately coincide with the end of the first quarter in October, extends from 12:30 P.M. to 10:30 P.M. Overnight liberty and long weekends may be authorized at the end of each quarter and for special events only at the discretion of the Commandant of Midshipmen.

In addition to having regularly scheduled weekend liberty, all midshipmen are given an opportunity for "dinner liberty" between the hours of 4:15 P.M. and 7 P.M. during the academic week. First classmen may request dinner liberty each weekday, and Plebes one day each week.

The Commandant may grant sick leave or emergency leave to a midshipman when circumstances warrant, and may also grant special leave or liberty for extracurricular activities and special events; a leave or liberty may be curtailed if a midshipman's conduct is unsatisfactory. The Superintendent may extend, reduce, or cancel leave or liberty when he deems that such action would be in the best interests of the Academy.

Varsity and Intramural Athletics

The Academy seeks to promote the growth of each midshipman as a "whole person" and is thus concerned with physical development as well as with development of character and intellect. Physical fitness and athletics are, therefore, an important part of Academy life.

Intercollegiate sports like football are popular at the King's Point campus of the Merchant Marine Academy.

The varsity athletic program is complete and comprehensive, but emphasis on intercollegiate competition is kept in harmony with the academic obligations of the midshipmen. The Academy strives to develop a healthy interest in athletics and to field teams that can win their share of victories over natural and friendly rivals. Such a program offers desirable development and recreation for the players and a healthy focus for midshipman loyalty. The Academy fields varsity teams in sixteen sports: baseball, basketball, bowling, crew, cross-country, golf, rifle, pistol, sailing, soccer, swimming, tennis, indoor track, outdoor track, football, and wrestling. Women can participate in all varsity sports except football and wrestling; however, in crew, women can participate only as coxswains.

The intramural athletic program offers all midshipmen the opportunity to enjoy the benefits of competitive sports at a level appropriate to their athletic ability. The program is organized around twenty different athletic activities, and is scheduled during the recreational hours. Competition is conducted on an individual, dual, and team basis, enabling each person to choose the type of activity best suited to his or her interests and capabilities. Competition between companies and battalions brings added zest to the intramural program, which includes the following sports: basketball, bowling, cross-country, handball, rowing, sailing, softball, swimming, table tennis, tennis, touch football, track and field, volleyball, wrestling, soccer, paddleball, rifle, pistol, badminton, and billiards. The interests of the midshipmen also extend beyond the intercollegiate athletic program, and club sports are offered for men in rugby, martial arts, fencing, weight lifting, gymnastics, and volleyball. Club sports are offered for women in fencing, gymnastics, and martial arts.

The Kings Point Sailing Squadron

Under the guidance of a professional sailing master, the Kings Point Sailing Squadron offers all interested midshipmen an opportunity to enjoy top competitive intercollegiate and ocean racing almost year 'round. A fleet of thirty interclub dinghies and fifteen 420s provides intercollegiate sailing at its best. A fleet of six thirty-foot Shields sloops, specially designed and donated by Cornelius Shields, participates in frequent races on Long Island Sound. Academy yachts, skippered by Academy midshipmen, have participated in the grueling Marblehead to Halifax race, the Annapolis to Newport and Bermuda races, and the Martha's Vineyard race. A collection of silver cups and trophies attests to the prowess of Academy sailors who have been nationally ranked in the top ten during the past few years. The Squadron also offers interested midshipmen an opportunity to get experience in both single-screw and twin-screw small-boat handling.

Extracurricular Clubs and Activities

Midshipman clubs and activities fall into almost as many categories as there are individual interests. For those interested in music or broadcasting, there are the Regimental Band, the Dance Band, the Glee Club, the Protestant Choir, and the Regimental Broadcast Unit.

Midshipmen with a literary bent can participate in publication activities, which include *Hear This,* a monthly newspaper; *Bearings,* the handbook of the Regiment of Midshipmen; and *Midships,* the

While the Merchant Marine Academy is academically demanding, there is also ample time to participate in a variety of extracurricular activities.

Dozens of clubs cater to the special interests of midshipmen at the Merchant Marine Academy. Recording clubs are a favorite.

yearbook of the graduating class. Also included in the communications arts is the Regimental Information Service, which desseminates information about the Regiment and the Academy to news services and the public.

One of the busiest midshipman activities is the Debate Council. The purpose of the Council is to develop skill at public speaking, and the most proficient members of the Council are selected to take part in debate tournaments, competing with leading colleges and universities in the United States and Canada.

A wide variety of other activities, including the International Relations Club, Marlinspike Club, Radio Club, Chess Club, Camera Club, and Trident Club for scuba diving and spear fishing enthusiasts, is available on campus. In addition, there are campus chapters of the Society of Naval Architects and Marine Engineers and the Marine Technological Society, which cater to the extracurricular academic and professional interests of midshipmen.

Land Hall, the midshipmen activities building, is the center of much of the extracurricular activity. Recently modernized and refurbished, the facilities include a snack bar, a billiards and pool room, club rooms, reading rooms and lounges. A full-time social hostess coordinates activities and assists with the planning of formal Regimental dances, frequent informal mixers, and an assortment of other events designed to enrich the social life of the midshipmen.

Cultural Activities

For midshipmen interested in the arts and world affairs, a stimulating series of lectures, plays, and concerts is presented on campus. In past years, the Academy has been host to Supreme Court Justice William O. Douglas, President Gerald Ford, Congresswoman Shirley Chisholm, and the world-famous astronomer Sir Bernard Lovell. Bowditch Auditorium has also been the scene of performances by pianists Peter Nero and Ferrante and Teicher, the Tommy Dorsey Band, the Trinidad Steel Band, Carlos Montoya, and the Juilliard String Quartet. In addition, the fact that the Academy is only twenty-three miles from New York City places midshipmen within each reach of the world's foremost cultural center, with its rock concerts, symphonies, Broadway shows, professional sports, museums, opera, and ballet.

Spiritual Life

Participation in religious activities and attendance at chapel services are strictly voluntary, being left to the needs and desires of the individual midshipmen. The U.S. Merchant Marine Memorial Chapel, built by public subscription as a tribute to the officers and men of the merchant marine who lost their lives in the service of the nation, serves all faiths.

A Protestant chaplain is available for counseling and guidance of

The Academy Memorial Chapel. The Chapel is for interfaith use by the midshipmen and staff. It is a memorial erected in memory of the men of the Merchant Marine who died at sea in wartime.

midshipmen. General Protestant worship is conducted on Sundays, and Holy Communion is held on the first Sunday of each month. Midweek service is at noon on Wednesdays. The Christian Council, Chapel Choir, Scripture Study class, and religious instruction provide other opportunities for religious activity.

A Catholic chaplain is also available daily for counseling, guidance, and religious instruction. The Blessed Sacraments are observed in the Chapel of Our Lady—Star of the Sea, and Mass is celebrated there each day at noon. The Newman Club and a weekly Theology Seminar add an extra dimension to the religious program, and a coed religious retreat weekend is held each spring. Before graduation a Pre-Cana Conference is conducted for those anticipating marriage soon after graduation.

Jewish services are conducted at the Academy, and Jewish personnel are also given special permission to attend one of the local synagogues. Arrangements are made for special leave and liberty for the observance of Yom Kippur, Rosh Hashanah, and Passover.

Midshipmen of other faiths requiring special arrangements to attend the church of their choice may obtain assistance from one of the chaplains.

THE U.S. NAVAL RESERVE MIDSHIPMAN PROGRAM

Relationship Between The Academy and The U.S. Navy

Primary responsibility for the merchant marine and the education of merchant marine officers is vested in the Secretary of Commerce, but a strong relationship exists between the Academy and the U.S. Navy. The Navy's interest in the Merchant Marine Academy stems from the national defense requirement for an adequate merchant marine, manned by well-trained officers who possess an understanding of naval procedures, so that merchant vessels are capable of operating with the Navy in time of war. The Navy gives support and assistance to promoting a strong merchant marine, fully recognizing the importance of the merchant marine to the United States. A vital part of that support is the Navy's participation in the training of midshipmen at the Academy.

Appointment of Candidates in the U.S. Naval Reserve

The Secretary of the Navy has authorized the appointment as Midshipman, USNR, those qualified midshipmen accepted for enrollment at the U.S. Merchant Marine Academy. Each applicant is screened by the Commander of the Navy Recruiting Command on the basis of physical qualifications, security clearance, and character. Each acceptable man or woman candidate will execute the acceptance and oath of office for appointment as Midshipman, USNR, during the month of September.

The Naval Science Program

All midshipmen are required to take a program of naval science courses administered by the Department of Naval Science at the Academy. The department is staffed by officers and men who are assigned to the Academy by the Department of the Navy. The Chief of Naval Education and Training prescribes the naval science curriculum and furnishes required textbooks, references, and training aids.

The performance of midshipmen is monitored by the Head of the Department of Naval Science. In the event a midshipman fails to display those qualities of leadership, character, and aptitude required of a prospective naval officer, the Head of the Department of Naval Science will make a recommendation to the Commander, Navy Recruiting Command, via the Superintendent of the Academy, that the midshipman's appointment be terminated and, upon termination of the appointment, the midshipman may be separated from the Academy.

Commission As Ensign, United States Naval Reserve

Upon graduation and Coast Guard licensing as a third mate or third assistant engineer, and with the recommendations of the Head of the Department of Naval Science and the Superintendent of the Academy, midshipmen are commissioned as ensigns, United States Naval Reserve, Inactive Duty, in the Inactive Duty Merchant Marine Program.

Service Obligation

After volunteering for and accepting appointment as an ensign in the United States Naval Reserve, Inactive Duty Merchant Marine Program, a graduate must perform satisfactorily in the Reserve for six years. During the initial years, a graduate has four service options:

1. Sail on his or her license aboard a U.S. flagship for not less than six months each year for the first three consecutive years immediately following acceptance of the commission, or
2. Sail on his or her license aboard a U.S. flagship for not less than four months for the first four consecutive years immediately following acceptance of the commission, or
3. Apply for and serve on full-time active duty as a commissioned officer in a uniformed service of the United States for a period of three consecutive years, or
4. Serve on active duty for training aboard a U.S. Navy ship for thirty consecutive days each year for three consecutive years immediately following acceptance of the commission, and be either employed ashore for the balance of each year in some phase of the maritime industry or engaged in full-time graduate studies related to the maritime field.

In addition, regardless of the option selected, graduates are required to complete at least one naval correspondence course per year for each of the six obligated years. Finally, they must accumulate twelve retirement points per year for each of the six years. These retirement points may be obtained by completing a correspondence course worth at least twelve retirement points.

THE ACADEMIC PROGRAM

The Merchant Marine Academy offers a four-year undergraduate program that leads to a Bachelor of Science degree and a merchant marine license as a third mate or third assistant engineer. In addition, graduates are commissioned as ensigns in the United States Naval Reserve. The Academy is accredited by the Middle States Association of Colleges and Secondary Schools.

Three major curriculums are offered: nautical science for the preparation of deck officers, marine engineering for students interested in becoming engineering officers, and a combination of the two, a dual license curriculum, which leads to a license in each specialty. In addition to a major, each midshipman may also take a minor or a concentrated elective program in such specialized fields as oceanog-

The Merchant Marine Academy's Department of Engineering offers excellent courses in all types of Naval and powerplant engineering.

raphy, nuclear engineering, management science, computer science, mathematics, chemistry, and naval architecture. General education courses make up about one-third of each of the professional curriculums, and all midshipmen are required to take naval science courses prescribed by the Department of the Navy.

The academic year at the Academy is divided into four academic quarters, which span eleven months, from the last week of July to the end of June. As an integral part of the academic program, midshipmen spend one-half of both their sophomore and junior years sailing on merchant vessels.

The curriculum at the Academy is thus stimulating and comprehensive. It is designed to assure that each midshipman, upon graduation, will be professionally competent, trained for leadership and responsibility, and well rounded intellectually.

In addition to their major, midshipmen may also pursue an academic minor in one of 15 specialized fields.

Each midshipman, during half of the sophomore year and half of the junior year, serves five months at sea on U.S. flag vessels.

Between periods of shipboard training during the sophomore and junior years, midshipmen return to the Academy to continue academic work.

147

Every midshipman takes courses in
nautical science and marine engineering.

Graduation Requirements

To be eligible for graduation, a midshipman must pass all required courses, earn the minimum number of credit hours prescribed for his curriculum, attain a cumulative quality point average of 2.00, pass the appropriate USCG license examination, and apply for, and accept if offered, a commission in the USNR.

The Program of Study—Plebe Year

All plebes follow a common program of study for the first two quarters of the freshman year. During this period, in addition to basic courses in mathematics, science, and the humanities, every midshipman takes introductory courses in nautical science and marine engineering. The new midshipman is thus given an opportunity to determine intelligently an area of special interest before choosing a major field of concentration.

The Shipboard Training Program

As part of the professional training, each midshipman spends two quarters of the sophomore year and two quarters of the junior year at sea, aboard commercially operated merchant ships. Every effort is made to assign midshipmen to several different vessels during their two periods of training. They thereby become familiar with the performance and operating characteristics of various classes of ships and with the diverse operating requirements of different trade routes, while at the same time gaining valuable practical experience in the performance of shipboard duties.

The most unique and exciting part of the Academy curriculum is the Shipboard Training program.

The shipboard training program provides all midshipmen with the opportunity to use a ship as a seagoing laboratory. Everyone is given a study guide called a "Sea Project" and, in addition to performing shipboard duties, is required to complete written assignments, which are submitted periodically to the Academy for evaluation and grading. The assignments are carefully designed to assure that, while aboard ship, midshipmen apply the knowledge and skills learned in the Academy classrooms and acquire a firm foundation for advanced study upon their return to the Academy. Written assignments cover the following areas:

Nautical Science	*Marine Engineering*
Navigation	Refrigeration and air conditioning
Seamanship	Diesel engineering
Cargo	Marine engineering operational systems
Rules	
Rules of the road	Electrical engineering
Ship construction	Ship construction/naval architecture

Shipboard training is the highmark of four years training at the U.S. Merchant Marine Academy.

Labor relations and ship's business	
Weather for mariners	Machine shop
Humanities	Humanities

Aboard ship, nautical science majors are assigned to the Deck Department and marine engineering majors to the Engineering Department, and Sea Project assignments concentrate on subject matter appropriate to the midshipman's specialty. However, during the first period of training, deck and engineering majors are required to complete assignments in the opposite department to assure basic familiarity with all aspects of ship operation. Dual license majors spend half their time at sea in each department, and their study program is designed to assure intensive experience in both specialties.

Should a student, prior to the second sailing period, acquire a definitive maritime career goal, a program of specialization can be arranged that will provide special opportunities for experience in the last sea period. Through counseling, the Head of the Department of Shipboard Training can assure the student of assignments that will provide maximum exposure in the desired area of specialization.

The sea year is concluded with a two-week assignment ashore for internship training in a maritime-related activity. Depending upon a midshipman's field of specialty and his individual interests, he or she may be assigned to a steamship company, shipyard, ship repair facility, ship brokerage/chartering firm, stevedoring firm, surveyor's office, towing company, port and terminal facility, or some similar enterprise. Each midshipman is required to complete a written report on these experiences, which is submitted to an appropriate department at the Academy for evaluation and grading.

Midshipmen receive a total of fifteen quarter-hours of credit for courses and reports completed as part of their shipboard training and internship during the sophomore and junior years.

Academy Training Representatives in New York, New Orleans, and San Francisco assign midshipmen to ships, monitor and guide their progress, and maintain liaison between the midshipmen, the Academy and the steamship companies. Fitness reports and progress reports are submitted at frequent intervals.

The Core Program of Professional and General Education

Between periods of shipboard training during the sophomore and junior years, each midshipman returns to the Academy and continues academic work in his chosen field. Each major has a core curriculum consisting of appropriate professional and technical courses in nautical science or marine engineering, and common courses in the sciences, humanities, and management areas. The dual license core contains courses in both nautical science and engineering.

A midshipman's senior year is devoted primarily to completion of the elective requirement, intensive study in the major, and preparation for the written examinations, administered by the United States

Coast Guard, that lead to licensing as a third mate or third assistant engineer.

Electives and Minors

Every midshipman, in addition to completing the required core curriculum in the major, is also required to complete a specific number of elective courses. The nautical science major must complete twenty-one quarter-credit-hours of electives, and the marine engineering major must complete eighteen quarter-credit-hours. Dual license majors are not required to complete an elective program since the dual license curriculum is, itself, an elective.

To meet the elective requirements, a midshipman may choose any course for which he has the prerequisites, or he may complete a prescribed sequence of elective courses leading to a minor in a specific academic discipline.

Minors normally consist of eighteen quarter-credit-hours (a few have higher credit requirements). The course sequence and credit-hour requirements are listed for each minor. A midshipman may complete more than one minor. However, not more than six quarter-credit-hours may be common to two different minors.

The following summary lists the minors:

Department of Engineering
  computer science
  marine electrical power and control
  marine machinery design
  marine thermal power systems and control
  naval architecture
  nuclear engineering
Department of Maritime Law and Economics
  law
  management
  maritime transportation
Department of Mathematics and Science
  chemistry
  mathematics
  physics
Department of Nautical Science
  marine electronics
  marine petroleum operations
  oceanography

Grades

The Academy uses a letter-grade system with each letter grade being assigned a numerical quality-point equivalent. The scholastic significance of the grades is as follows: A is outstanding; B is above average; C is average; D is minimum passing grade; F is failure; I is incomplete. A plus is appended to the letter grade to indicate that a midshipman is near the top of the grade category.

Quality-point values per quarter-credit-hour are assigned to letter grades in accordance with the following: A = 4.00 to D = 1.00.

Promotion and Graduation

A midshipman is considered to be in satisfactory academic standing, eligible for promotion and graduation, if he or she has met the specific requirements of the Academy's academic policies, which are in general: pass all required courses and have a cumulative quality-point average of at least 2.00. Fourth classmen are permitted an adjustment period during which they may fall below 2.00; however, they must meet other specific minimums approved by the Superintendent.

CAREERS IN THE MARITIME INDUSTRY

Young men and women graduates of the Academy can find rewarding careers in the exciting, worldwide maritime industry.

Merchant Marine Officers

Merchant vessels are manned by deck and engineering officers who are licensed for such service by the U.S. Coast Guard, the federal agency responsible for maritime safety. A license is issued only after a candidate has met certain minimum requirements of training or experience and has established competence by passing a comprehensive written examination covering a broad range of professional subjects. Deck licenses, in ascending order, are for third mate, second mate, chief mate, and master. Engine licenses are for third assistant engineer, second assistant engineer, first assistant engineer, and chief engineer.

Deck officers are responsible for the safe navigation of the vessel, loading and discharge of cargo, vessel maintenance, and shipboard safety. Engineering officers are responsible for the maintenance and operation of all machinery aboard ship, including propulsion machinery, auxiliary machinery, electrical systems, refrigeration machinery and air-conditioning systems.

After graduation from the Academy, a newly licensed third mate or third assistant engineer joins a ship as a fully qualified junior officer. He or she will be in charge of a watch, on the bridge or in the engine room, and will be responsible for the safety of the ship or its propulsion plant while on duty. After a year of shipboard experience, the examination for a second mate's or second engineer's license may be taken; then after serving a year as "Second," the young officer is qualified to sit for the chief mate's or first assistant engineer's license. The final step, after additional service and experience, is to sit for the license of master or chief engineer.

Merchant marine officers are usually members of maritime labor unions, but are employed by private steamship companies in accordance with negotiated collective bargaining agreements.

Salaries for merchant marine officers are excellent, and union pension plans make it possible to retire with a pension after twenty years of service.

Modern, high-speed merchant ships are large and complex. Their sophisticated power plants and navigational systems require the services of well-trained, highly professional officers, and a career at sea can be interesting, challenging, and rewarding. The availability of positions at sea for the new graduate naturally varies from year to year.

A whole new area of shipboard employment is opening for Academy graduates in domestic shipping on the Great Lakes and the nation's major rivers, as well as in the rapidly growing offshore mineral and oil industry.

Marine Transportation Management

The future growth and productivity of the nation's merchant marine depends not only on a modern fleet of ships, but also on the expertise and quality of management in all sectors of the maritime industry. Accordingly, the Academy graduate, after service at sea, will find challenge and opportunity ashore.

In the marine department of a steamship company the marine superintendent, usually a senior master mariner, is responsible for all matters concerning the deck departments of the company's vessels plus the operation and maintenance of terminal facilities. A competent marine superintendent has usually worked up through the posts of ship's officer, shipmaster, and port captain, and has acquired a knowledge of naval architecture and ship repair, terminal operations, stevedoring, marine insurance, and labor relations. Superintendents are assisted in the performance of their duties by subordinates known as "port captains," who are also experienced ship's officers or masters.

The engineering counterparts of the marine superintendents are the superintendent engineers, who are responsible for all marine engineering activities of a company's fleet. They and their assistants, the "port engineers," are experienced engineering officers who, after service at sea, have been promoted to posts ashore because of their technical and administrative abilities.

Opportunities ashore are not limited, however, to masters and chief engineers. Other Academy graduates, with varying degrees of seagoing experience, find positions ashore with steamship companies. In addition, Academy graduates are employed in a wide variety of related fields: stevedoring companies, which are responsible for the loading and unloading of ships' cargoes; marine insurance firms, which handle the insurance and settlement of claims for shipowners, ship operators, exporters, and importers; ship brokers and chartering agents, who arrange contracts for the employment of ships and the carriage of cargoes; municipal and state port authorities; and companies that operate tugs and barges are but a few of the enterprises that offer rewarding and challenging careers to graduates interested in transportation management. A young person interested in law might find admiralty law an interesting field after completing postgraduate education at a law school. The Academy's minors programs in management, law, and transportation provide undergraduate preparation for entry positions in maritime management and graduate business study. As a result, Academy graduates have acquired advanced degrees in business or law from such schools as NYU, Harvard, Yale, Columbia, Pennsylvania, Northwestern, Michigan, and UCLA, among others, and many occupy top management positions in the maritime industry.

Naval Architecture and Marine Engineering

A ship is virtually a floating city. Every ship must contain its own kitchen facilities, sanitary facilities, laundry, recreational, and sleeping facilities. Every service necessary for the people of a city must be provided for the ship's crew and passengers, and precisely designed units, down to reading lamps, must use every cubic inch of space efficiently. The resulting complex of systems and facilities must also be capable of plowing through surging seas, whose powerful forces continuously batter the ship from every direction.

The naval architect and marine engineer must be prepared to employ both art and science in the design and construction of a ship and its machinery. Using imagination and experience, the architect must translate a shipowner's functional requirements into a suitable and economic design of appropriate size, hull form, and speed, and the marine engineer must design, test, and install a compatible power plant. Naval architects and marine engineers are also called upon for the conversion and repair of existing ships, and actively engage in research and development activities related to their specialties.

The problems related to the design, construction, repair, and conversion of ships and propulsion machinery are so complex that a wide variety of skills and disciplines are involved. For many career opportunities, the marine engineering curriculum at the Academy, supplemented by experience at sea or on-the-job training and experience, is adequate preparation for placement and advancement. However, many specialties within the profession require graduate education in marine, mechanical, or electrical engineering or in naval architecture and design.

Oceanography

Oceanography is the scientific study of all aspects of the ocean—its boundaries, characteristics, movements, physical properties, and life. Oceanographers not only extend the basic scientific knowledge about the ocean, but, in addition, contribute to the development of practical methods for fully exploiting the ocean and its resources.

Most oceanographers are specialists in one of the branches of the profession, such as biological, physical, geological, or chemical oceanography, and most work part of the time aboard oceanographic ships at sea on voyages that may last from a few days to several months. A few oceanographers work nearly all the time aboard ship, while some do not go to sea at all, but pursue mathematical or theoretical studies ashore.

Only a few colleges offer undergraduate programs in oceanography. However, since oceanography is an interdisciplinary field, training in one of the basic related sciences, when coupled with a strong interest in oceanography, is adequate preparation for entry into graduate school or for some beginning positions in the field. The basic curriculums at the Academy, supplemented by the elective program in oceanography and postgraduate study, will provide an Academy graduate with an excellent foundation upon which to build a career in oceanography.

Military Career Opportunities

As an ensign in the United States Naval Reserve, the Academy graduate has an opportunity to apply for and serve on active duty in the Navy. Many graduates, after serving on active duty and finding the Navy both challenging and rewarding, have remained in the service as career officers. Today, Kings Pointers are serving in the navy in surface ships, in submarines, and as aviators, in every rank from ensign to rear admiral.

Similar career opportunities are available in the United States Coast Guard and in the National Oceanographic and Atmospheric Administration (NOAA).

Academy Alumni

The list of distinguished Academy alumni is long and extensive, representing a broad spectrum of professional accomplishments. Among the notable graduates are Captain John Clark, '40, president of Delta Steamship Lines; Erik F. Johnsen, '45, president of Central Gulf Steamship Corporation; Edward Heine, '46, president of United States Lines, Incorporated; Fred S. Sherman, '55, president of Calmar Steamship Corporation; Dr. William B. Morgan, '50, Supervisor Naval Architect of the David Taylor Model Basin; Dr. Joseph M. Chamberlain, '45, Director of the Adler Planetarium; Captain Stuart F. Sammis, '50, Chief Surveyor of the National Cargo Bureau; Carroll N. Bjornson, '50, president of Stolt-Neilsen Chartering; Rear Admiral Carl J. Sieberlich, USN, '43, and the late Elliot M. See, '49, one of the first civilian astronauts. The Academy is proud of the record made by its graduates: afloat and ashore, in private industry and public service, in civilian and military careers, Kings Pointers serve their nation well.

QUALIFICATIONS FOR ADMISSION

General Requirements

All candidates must meet certain requirements of citizenship, age, and moral character, but the Academy considers qualified applicants without regard to race, color, sex, or national or ethnic origin.

Citizenship. All candidates must be citizens of the United States, either by birth or by naturalization.

Naval Reserve Midshipman Requirement. Candidates must meet the physical, security, and character requirements for appointment as Midshipman, USNR.

Nominations

A young man or woman who desires to enter the Academy upon graduation from high school must be nominated by a member of the Congress of the United States or another nominating authority. Senators and Representatives of the United States Congress; the governors of the Canal Zone, Guam, American Samoa, and the Virgin Islands; the Resident Commissioner of Puerto Rico; and the mayor of the District of Columbia may each nominate ten candidates for the Academy (the governor of the Canal Zone can nominate males only). Special legislative provisions permit the appointment of midshipmen from other American republics and from the Pacific Trust Territories.

Nominating officials select their nominees by any methods they wish, including a screening examination. This examination may be given as early as July of the year before appointment is sought. The test is for nomination and is not to be confused with the examinations required for appointment.

A candidate should apply for a nomination early. Some nominating officials establish deadlines for the receipt of nomination requests in order to allow adequate time for processing and evaluating requests. The best time for a candidate to apply for a nomination is in May of his junior year in high school.

Nominating officials may submit the names of their nominees to the Maritime Administration any time between August 1 and December 31 of the school year preceding that in which admission to the Academy is desired.

Requesting Nomination

You may request nomination as a candidate for admission to the Merchant Marine Academy by writing a letter in the form shown in the back of this book and addressing it to an appropriate nominating authority. Be sure your full name is typed or printed legibly. A typewritten letter is preferred.

Application At the time of requesting nomination, or as soon as possible thereafter, a candidate should obtain an application from the Academy. Candidates who do not obtain an application will automatically be mailed one when their official nomination is received by the Admissions Office, but filing an application will permit more rapid processing of a candidate's admission file.

Admission Test Requirements All candidates are required to take either the College Entrance Examination Board's Scholastic Aptitude Test (SAT) or the American College Testing Program's test (ACT) on scheduled dates at convenient testing centers throughout the country. Required testing must be completed by the first test date of the year in which admission is sought, and all tests should be taken within sixteen months prior to the month of admission.

Candidates must request the testing agency to submit their test scores to the United States Merchant Marine Academy, Kings Point, New York 11024. The cost of the examination must be borne by the individual candidate. The Academy's College Board code number is 2923; its ACT number is 2974.

Physical Requirements All candidates for admission to the Academy must be in good physical condition, and must meet the requirements for appointment as a Midshipman, USNR. Physical examinations are conducted by a service academy examining facility designated by the Department of Defense Medical Review Board, and a final decision on a cadidate's physical qualifications is made by that Board.

Nonswimmers Fourth classmen must demonstrate the capability of swimming 100 yards using two basic strokes and 15 minutes of flotation. Nonswimmers must devote a part of their study time to learning basic swimming strokes, and it is, therefore, recommended that applicants learn to swim before entering the Academy. This requirement must be fulfilled prior to the first shipboard training period.

Waivers No waivers of the general, scholastic, or physical requirements will be granted.

Candidates are appointed competitively by the Academy for the vacancies located in their state or geographical subdivision. Each state has a quota proportionate to the state's representation in Congress. After the principal appointees have been selected, the remaining qualified candidates will be designated as alternates, to be appointed in order of merit should vacancies occur within their states.

In the event there are insufficient candidates within a state to fill the state's quota, appointments to fill the unmet quota are made from the national list of alternates, ranked in order of merit as described above.

A candidate's competitive ranking is based upon performance on the required College Entrance Examination Board or American College Testing Program tests, high school class rank, leadership potential as demonstrated by participation in high school extracurricular activities, and employment experience.

Armed Forces Enlisted Personnel

Enlisted personnel on active duty and enlisted reservists on inactive duty in any of the armed services may request a nomination and compete for admission to the Merchant Marine Academy in the same manner as other candidates. Enlisted men or women who are selected for appointment will be released by their service to accept orders to the Academy in accordance with the following conditions:

Merchant Marine Academy graduates look forward to interesting and rewarding careers in the Maritime service as commissioned officers.

158

1. They will not be discharged from enlisted status by reason of the appointment.
2. They will hold dual status as midshipmen and as enlisted men or women while attending the Academy, but will accrue longevity while in this dual status.
3. When appointed as commissioned officers upon graduation, they will be discharged as enlisted men or women.
4. A former enlisted man or woman who is disenrolled from the Academy prior to graduation will have midshipman status terminated and may be recalled to duty.

Service Obligation Agreement

All midshipmen sign an agreement as a condition of admission that obligates them to serve in one of the four ways after graduation as described earlier.

ALLOWANCES AND EXPENSES

The major cost of attendance at the Academy is borne by the government. There is no tuition fee, and midshipmen are provided with comfortable quarters and three well-balanced meals each day. Medical and dental care is provided at the Academy infirmary through the United States Public Health Service.

Midshipmen receive a government allowance toward the cost uniforms and textbooks that currently totals $1725.00 over the four-year course. It may be used only for the purchase of prescribed uniforms and textbooks and, on the average, is usually sufficient for these purposes. However, this may not be so in individual cases, when it becomes the responsibility of the midshipman and his parents to offset any deficiency. Shortages usually occur when midshipmen do not complete the full course. In such cases they receive only a pro rata share of the annual governmental allowance, and are responsible for expenses incurred over that amount. While assigned aboard ship, midshipmen are provided with quarters, meals, medical care, and are paid $333.00 per month.

The Academy publishes a list of the required items for which government funds are provided. In order to assure uniformity of appearance, quality, and cost, all required items are purchased by an authorized Academy staff member on specifications approved by the Commandant of Midshipmen or the Dean as the case may be. Second-hand items are not acceptable in lieu of original purchases, nor are second-hand replacement items acceptable. The government allowance is provided solely for the required uniforms and textbooks and must be expended only for the items so listed.

Cash deposits are required from midshipmen to pay for personal items of clothing, school supplies, and such services as laundry, haircuts, tailoring, movies, cultural programs, athletic events, yearbook, and miscellaneous personal expenses. These deposits are maintained and accounted for individually and may be used only for the stipulated expenses. Any excess is to be returned upon leaving the Academy. Deficits must be offset by additional deposits. In the event that a deficit occurs in a midshipman's government account, funds to offset the deficit may be transferred from a personal deposit account. However the reverse does not hold.

The following deposits are required:

plebe year (freshman): deposit prior to reporting, mailed by June 30	$800.00
third class (sophomore)	$225.00
second class (junior)	$225.00
first class (senior)	$225.00

Both the government and personal accounts are maintained by the government, and computer printouts showing all transactions and balances are furnished to the midshipman monthly.

Expenses for spending money and transportation during leave periods are not included in the amounts mentioned above.

Failure to make the required deposits may result in refusal of enrollment, suspension or detachment.

Student Loans

It is recognized that some candidates may require financial assistance in order to obtain the required deposit. Since the Academy can offer no direct financial aid other than the free tuition, room, board, and government allowances, candidates are urged to act early and take the necessary steps to avail themselves of financial assistance offered by local banks and state loan programs. It is recommended that candidates investigate the availability of such loans during the fall of their senior year in high school, seeking advice from bank officers and high school guidance counselors if necessary. If after applying for a local bank and/or state loan, a candidate is unable to obtain the required financial assistance, he or she should communicate with the Admissions Office, U.S. Merchant Marine Academy, Kings Point, New York 11024.

Health and Accident Protection

Medical and dental care is provided without cost to midshipmen by the Academy or by the United States Public Health Service. Injuries that occur while on duty are covered by the Bureau of Employees' Compensation of the United States Department of Labor. Such treatment may continue, if required, after graduation or separation. In the event of disability as the result of an accident, the Bureau of Employees' Compensation will award a cash settlement.

Eye glasses are not furnished. If a student needs to have glasses repaired or to purchase new ones, this expense must be borne by the individual.

The government does not provide life insurance for midshipmen. Parents are expected to provide such insurance through individual or group plan coverage available from private insurance companies.

Questions and Answers About the Merchant Marine Academy

What do I have to do to attend the United States Merchant Marine Academy?

All candidates must: be nominated by a member of Congress, or another nominating authority; be academically and physically qualified; compete well for a vacancy from the state or locality of nomination; and meet the standards for an appointment as Midshipman, USNR.

Are the requirements for women different from those of men?

Basically, no, except for standards such as height and weight. They compete for appointment, but the same academic selection criteria are used.

Women must meet the same admissions selection criteria as men to enter the Merchant Marine Academy.

Does one have to have political "pull" or know someone to get a congressional nomination?

No, not at all. Each congressman or nominating authority has guidelines and/or a selection board whose requirements usually conform to those of the Academy.

When should I apply for a nomination?

Preferably in May of your junior year. However, requirements differ from one congressman to another, so check with your Representative and the two U.S. Senators from your state.

Who makes the final selections or offers the appointments?

The Academy.

How?

On the basis of the candidates's: SAT or ACT results; overall high school or academic record; class rank; and motivation toward a maritime career, interest in the Academy, industriousness, citizenship, and recommendations from high school counselors, teachers and/or principal.

Then, a candidate for the Academy may receive a nomination, but this in no way assures him or her of gaining an appointment?

That's true. Each nominee must first compete with other nominees from his or her state.

Suppose I don't compete well with the other nominees from my state, do I have another chance?

Yes. All candidates are ranked in order of merit, based upon the criteria above. They compete first within the state from which they are nominated; then they may compete for any unfilled national vacancies. For example, New York State has a quota of twenty-two; should it have only eighteen qualified candidates, four vacancies would be placed in the national pool, and candidates could compete, in order of merit, for those vacancies and any others that might occur in other states.

What should I do if I'm too late or am unsuccessful in obtaining a nomination from my senator or representative?

Your Senator or Representative may suggest an alternate nominating authority or you may seek a nomination from a congressman in your state who did not submit the allowable ten nominations.

What's the latest date my application can be sent in?

The deadline is March 1. However, if, due to extenuating circumstances, you were nominated late or did not receive your application before that date, you may send your application in and possibly compete for a national vacancy.

Are there certain required courses I must have taken in high school?

Yes. You must have taken at least three years of English, three years of mathematics (algebra, geometry, and trigonometry), and one year of chemistry or physics with a laboratory.

Is there an age requirement?

Yes. Candidates must be at least seventeen and not have passed their twenty-second birthday on July 1 of the year of admission. However, a waiver may be granted for veterans of the armed forces on the basis of one month for every month of service up to the age of twenty-four.

How can I tell what my chances of acceptance are?

Last year, the average candidate had a SAT verbal score of 510, a mathematics score of 600, and ranked in the top 20 percent of his or her class.

What are the minimum academic requirements of the Academy?

A candidate should be in the top 40 percent of his or her class and have a minimum SAT composite score of 950, with a score of at least 500 in mathematics; or a candidate should earn scores of at least 21 in English and 25 in mathematics on the ACT.

What if I'm in a highly competitive school and am not in the top 40 percent?

Your test scores and high school average will be the more determinate factors.

If I am designated a principal candidate, am I assured of getting in?

No, not necessarily. You must satisfactorily pass the medical, security, and Midshipman, USNR requirements.

Is a physical aptitude test used at the Academy?

No.

Will an NROTC physical satisfy the Academy's requirements?

The four-year NROTC physical will, but the two-year one will not.

What are the visual requirements?

Visual acuity in each eye must be 20/100 uncorrected, or better. Each eye must be correctable to 20/20. Refractive error must not be greater than plus or minus 3.25.

Can women compete in all sixteen varsity sports?

All except two—football and wrestling.

Do women attend the same classes as men?

Yes.

Do women also participate in the shipboard training program in which midshipmen live on merchant ships during one-half of each of their sophomore and junior years?

Yes.

Am I required to serve on active duty in the armed services upon graduation?

No. Graduates are trained principally to be officers in the merchant marine and leaders in the maritime industry. However, one optional way of fulfilling your service obligation is to go on active duty as an officer in the U.S. Navy or Coast Guard.

Will women have the same career opportunities as men when they graduate?

The Academy anticipates that there will be no difference. Women graduates will have had the same training as men and will be capable of participating in the same career areas, with the possible exception of some jobs in the offshore drilling area.

6

The State Maritime Academies

The Massachusetts Maritime Academy

The Maritime College of the State University of New York

The California Maritime Academy

The Texas Maritime Academy—Moody College of Marine Sciences and Maritime Resources—Texas A & M University

The Maine Maritime Academy

The Great Lakes Maritime Academy

Federal authority and encouragement for developing state maritime academies date from an Act of Congress of 1874. Over the years since then, six states have created educational institutions to help provide the nation with the Merchant Marine force needed to support America's domestic and foreign commerce, and to meet the requirements for national defense. The states, in operating the maritime academies, obtain considerable support from several federal agencies: U.S. Maritime Administration, U.S. Navy, U.S. Coast Guard, and the U.S. Public Health Service.

The U.S. Maritime Administration's interest stems directly from a mandate of congress, expressed in the Merchant Marine Act of 1936, which directs the maintenance of an adequate Merchant Marine to support the country's maritime requirements and responsibilities on the oceans and waterways around the world, and on the coastal and seaways of this nation.

The six states that have maritime academies are California, Maine, Texas, New York, Massachusetts, and Michigan.

THE MARITIME INDUSTRY

The maritime industry is an economic activity that includes the seagoing operations of the merchant marine and shore activities such as shipbuilding and repair, port operations, marine insurance, admiralty law, transportation industry management and operations, ship brokerage and chartering, equipment design and manufacture, waterborne foreign and domestic commerce, and all aspects of ocean resources. In addition to serving in such professional capacities as engineering, science, and business personnel in the maritime industry, graduates also serve in many related fields such as public utilities, nuclear energy research, development, and operations, and environmental protection, to name but a few.

National recognition of the problems of dependence on foreign carriers for materials and energy, and the American genius at technological advancement are setting the stage for projecting the American merchant marine to a more prominent position in world commerce. The need for highly trained personnel as officers on American ships is increasing at an even faster pace as a result of the planned retirement of many senior officers, the masters, chief mates, chief engineers and first engineers. In a recent manpower study, a government agency projected excellent opportunities for entering employees and predicted rapid advancement to develop well into the next decade. It is expected that in the critical years of 1979 through 1985 the Maritime College graduates will be a very important source to meet any projected demand for new employees.

The complexities of international business and the advances of technology have not diminished the rewards of the sea experience. Seagoing work offers opportunities, literally, to see the world, to enjoy the camaraderie of seafaring men of all nations, and to assume positions of responsibility that are unique. As mankind looks to the oceans as a frontier of natural resources, highly skilled and dedicated personnel will find challenges and personal rewards.

In addition to the personal rewards that attract men to the sea, the financial rewards are exceptional. Attractive starting salaries for recent graduates "sailing on their license" are supplemented by liberal vacation benefits. Many officers make the sea their career, and others prefer to work in shore positions of responsibility in the maritime industry and related activities. Nearly all graduates of recent years have had early offers of appropriate professional employment.

U.S. NAVY OFFICER PROGRAMS

Graduate of Merchant Marine Academy Programs

Maritime Administration General Order 87, which specifies certain regulations and standards for maritime academies, requires that merchant marine license program students shall agree in writing to apply, at an appropriate time prior to graduation, for a commission as ensign in the U.S. Naval Reserve (Inactive), and to accept such commission if tendered. To assist students in the fulfillment of this obligation, the Navy has developed the Graduates of Merchant Marine Academies Program. While all seniors in the license program (except NROTC students) are processed for commissioning under this program, only those students who successfully complete all required courses in naval science are recommended for commissioning. The Training and Service Agreement for Inactive Duty Merchant Marine Appointees requires fulfillment of one of the following options by those students selected for commissioning upon graduation.

1. Sail on their license at sea for not less than six months each year for three consecutive years; or
2. Sail on their license at sea for not less than four months each year for four consecutive years; or
3. Apply for and serve on active duty for training on board a navy ship for a minimum period of thirty consecutive days each year, for a period of three consecutive years immediately following acceptance of commission; or
4. Apply for and serve on full time active duty in the Navy Service for three consecutive years.

In addition the appointee must:

1. Maintain an active status in the Naval Reserve by earning at least twelve retirement points each year, and
2. Apply for and satisfactorily complete at least one correspondence course each year.

As the Navy does not consider this program a significant source of active duty officers, only an extremely small percentage of each graduating class is offered orders to active duty and therefore may discontinue the program at some academies. Students interested in active duty with the Navy, or contemplating a naval career, are encouraged to apply for one of the NROTC programs described below.

Naval Reserve Officers Training Corps Program

The Naval ROTC program is designed to train and educate qualified students for ultimate commissioning and active service as officers in the U.S. Navy or Marine Corps. In order to be eligible for application for this program a student must:

1. Be a U.S. citizen;
2. Be at least seventeen but less than twenty-one years of age;
3. Be physically qualified;
4. Possess satisfactory records of academic ability and moral integrity;
5. Demonstrate those characteristics desired of a naval officer; and
6. Have no moral obligation or personal conviction that will prevent him from bearing arms.

The NROTC Scholarship Program offers the following benefits: all tuition paid, all books furnished, $100 per month subsistence allowance during the school year, and a substantial uniform allowance. Graduates of this program receive regular commissions in the United States Navy or Marine Corps, and are required to serve on active duty for four years. High school students may apply for the scholarship program from March 1 of their junior year to November 15 of their senior year. Application forms are available from any Navy recruiter, the Academy NROTC Unit, and most guidance counselors. Early application is recommended, as this program is highly competitive.

The NROTC College Program offers students not selected to receive a scholarship an opportunity to participate in NROTC. The monetary benefits of the college program include: a substantial uniform allowance and $100 per month subsistence allowance during the junior and senior years. Graduates of the college program receive reserve commissions, and are required to serve on active duty for three years. Academy students may apply for the college program from the beginning of the freshman year through the end of the sophomore year. For further information concerning either program contact your local Navy recruiter.

Shortly after World War I, in accordance with the provisions of a newly-enacted basic Naval Reserve Law, students of the state maritime institutions were accorded further Federal recognition, when the Navy conferred upon them the status and rank of Cadet, Merchant Marine Reserve. The Navy introduced into the curriculum a course in naval science and assigned officers of the Navy as instructors. By Executive Order of the President in 1941, Federal cognizance and responsibility concerning State Maritime Academies was transferred from the Navy to the U.S. Maritime Commission.

U.S. COAST GUARD

The United States Coast Guard has offered direct commissions in the Coast Guard Reserve to graduates of the Maritime Academies. Senior cadets may apply for a commission as ensign, USCGR, to become effective upon graduation. The Coast Guard looks to the Maritime Academies as a source of newly-commissioned officers to serve in the specialty of Marine Inspection Officers in the Coast Guard Merchant Marine Safety Program. Initial contracts are for three years active duty in the Coast Guard Reserve.

LICENSE REQUIREMENTS

In order to be eligible for the license as third officer in the Merchant Marine and, in turn, eligible for the federal subsidy while a student, students must meet requirements established by the Maritime Administration and U.S. Coast Guard. Students who do not meet the below-listed requirements by reason of citizenship or health, may attend the Maritime schools, but will not be eligible to receive the license or the federal subsidy while a student. Students who are fully qualified for an original license will be designated cadet, maritime service (USMS), and will be entitled to the federal subsidy and health care provided by the Public Health Service while a student. Only U.S. citizens are eligible for license as officers in the American merchant marine and, in turn, to receive the federal subsidy. Students who are not U.S. citizens or who are not physically qualified for licensure when they complete the license examination may receive certification by the Coast Guard that they meet all requirements for license with the appropriate exception.

Health

Applicants are required to complete a physical examination by a physician of their choice and at their expense. Medical forms to be used for this examination will be sent to the applicant by the schools. Upon completion of the entire physical examination, forms are to be returned to the Admissions office. Applicants who have applied for ROTC scholarship competition, or for admission to a service academy may request the Department of Defense Medical Review Board to forward a copy of their complete physical examination report to the Maritime academies.

In general, recognized illness or physical defects that would render the applicant incapable of performing the regular duties or would interfere with the ordinary duties of an officer at sea are disqualifying. Examples are epilepsy, diabetes, etc.

Coast Guard regulations concerning the licensing of Merchant Marine officers require the applicant to have uncorrected vision of at least 20/100 in both eyes, correctable to at least 20/20 in one eye and 20/40 in the other eye for deck officers; and at least 20/100 in both eyes, correctable to at least 20/30 in one eye and 20/50 in the other for engineering officers. The color sense of all applicants will be tested through the physical examination. Inadequate color perception may disqualify an applicant for licensure.

Age Requirements

Cadets must be eligible to graduate and be commisioned in the Naval Reserve before reaching their twenty-ninth birthday. The maximum age limit may be waived for veterans.

The Massachusetts Maritime Academy

The Massachusetts Maritime Academy has moved from an enrollment of 270 in 1972 to 810 in September 1975, a threefold increase. The professional faculty has experienced a corresponding jump in number from twenty-nine in 1971 to eighty-eight in 1975. These tremendous increases were designed to complement the new, $17 million fifty-five acre land campus, which opened its doors in January 1972. Because of its unique educational program and the lure of the sea, the Massachusetts Maritime Academy has always held a strong attraction for high school graduates. Now, however, entrance to this Academy is more appealing and more desirable than ever before.

The four-year academic program leading to Bachelor of Science degrees in marine transportation and marine engineering has recently received full accreditation from the New England Association of Schools and Colleges.

A new dormitory complex was completed in 1976, and the entire cadet corps is housed in an ideal two-to-a-room occupancy.

The new land campus, with its spacious gymnasium, baseball field, football stadium, and other outside facilities, has permitted the Academy to establish for the first time in its eighty-five year history a formal program of intercollegiate athletics. Recently organized Academy teams representing just about all sports, have quickly matured and are now ready to explode across the spectrum of New England small college athletics. The young athletes who enter this school during the next few years will always glory in their participation in what is going to be a spectatular era of athletic competition.

Job placement for the Academy's 1977 graduates was 100 percent effective; the average starting salary was an unbelievably high $18,000 plus and the outlook appears brighter.

The newly acquired training ship is in excellent condition ready to embark on fascinating world-wide training cruises. A highly specialized, enthusiastic, and talented faculty is eagerly waiting to educate and to develop young men into premier mariners.

HISTORY OF THE ACADEMY

The Massachusetts Maritime Academy, originally known as the Massachusetts Nautical Training School, was founded in 1891 as the result of an act passed by the Massachusetts Legislature. It is the oldest, continuously-operating maritime academy in the country. The school, whose purpose was, and is, to train young men to become officers in the U.S. Merchant Marine, was placed under the management and supervision of three Commissioners appointed by the Governor. The USS *Enterprise* was loaned as a training ship to the Commonwealth in 1892 by the federal government, and the first class of forty cadets was received on board in 1893. On April 13, 1895, the first class of twenty cadets was graduated, eight in seamanship and twelve in marine engineering.

In 1909 the USS *Ranger* replaced the USS *Enterprise* as the school's training ship and, although the *Ranger* was taken over by the Navy during World War I, it was permitted to remain under the immediate direction of the appointment Commissioners. The name of the *Ranger* was subsequently changed to the *Nantucket,* and still later to the *Bay State.* She was returned to the federal government in 1942. The ship was then recommissioned and renamed the *Emery Rice* in tribute to Captain Emery Rice of the Class of 1897, who, in command of the transport Mongolia, sank the first German submarine in World War I.

In 1946, the Massachusetts Maritime Academy was authorized to grant the Bachelor of Science degree. The first degrees were awarded to the Class of 1949.

From the date of founding until 1942, the school was located in Boston. In that year the name of the school was changed to Massachusetts Maritime Academy, and its location was transferred to Hyannis on property that was formerly the home of Hyannis State Teachers College.

All facilities of the Academy were transferred from Hyannis to its present location in Buzzards Bay adjacent to the Cape Cod Canal in 1948, at which time the ex-Navy gunboat USS *Charleston* became the training ship. In 1957 the U.S. Maritime Administration furnished on loan the USS *Doyen,* renamed the *Bay State,* as the training vessel.

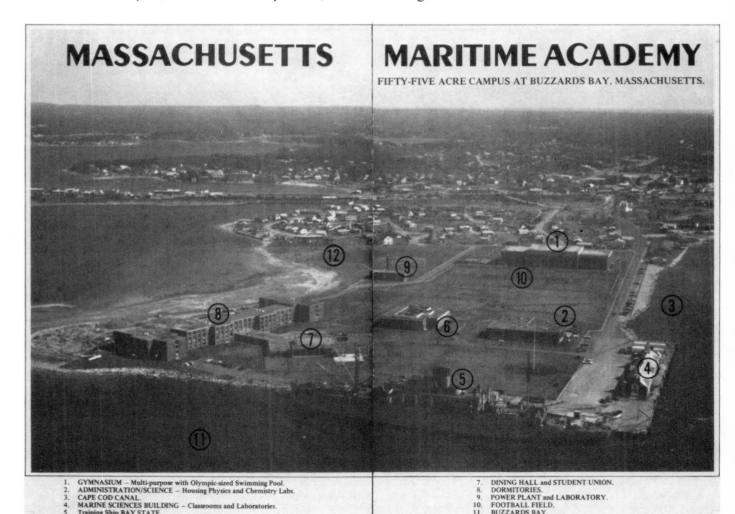

MASSACHUSETTS | MARITIME ACADEMY
FIFTY-FIVE ACRE CAMPUS AT BUZZARDS BAY, MASSACHUSETTS.

1. GYMNASIUM – Multi-purpose with Olympic-sized Swimming Pool.
2. ADMINISTRATION/SCIENCE – Housing Physics and Chemistry Labs.
3. CAPE COD CANAL.
4. MARINE SCIENCES BUILDING – Classrooms and Laboratories.
5. Training Ship BAY STATE.
6. Air-conditioned LIBRARY and Computer Center.
7. DINING HALL and STUDENT UNION.
8. DORMITORIES.
9. POWER PLANT and LABORATORY.
10. FOOTBALL FIELD.
11. BUZZARDS BAY.
12. BASEBALL FIELD

This training ship earned six battle stars as an attack transport in the Pacific during World War II. The *Bay State* was replaced in October 1973 by the former TS *Empire State* and renamed *Bay State*.

On June 18, 1964, new legislation placed the Massachusetts Maritime Academy within the Massachusetts State College System.

ACCREDITATION

The Massachusetts Maritime Academy is one of eleven colleges that comprise the state college system of the Commonwealth of Massachusetts. The Academy is authorized to grant the degree of Bachelor of Science in marine engineering and in marine transportation.

The U.S. Department of Health, Education, and Welfare, Office of Education, has certified the Massachusetts Maritime Academy as an eligible institution for various Federal grants and loans.

The Academy is accredited with the New England Association of Schools and Colleges.

The curriculums satisfy all Federal requirements for licensing by the U.S. Coast Guard as a deck or engineering merchant marine officer, and for a commission as ensign, USNR in the Merchant Marine Reserve 1105 Program.

Location of Massachusetts Maritime Academy.

LOCATION

The Massachusetts Maritime Academy is located in the village of Buzzards Bay, Bourne, in the southeast of the state, on a peninsula adjacent to the Cape Cod Canal. It is conveniently accessible by car from all land points via highways 24 and 25 and 195. Interstate public bus transportation also serves the area.

CAREER COUNSELING AND PLACEMENT

Since 1938 the Navy has actively participated in the training of maritime officers by teaching naval science courses.

Specific Navy interest in maritime training schools stems from the national defense requirement for an adequate merchant marine, manned by well-trained officers possessing an understanding of naval procedures, and capable of operating jointly with the Navy in time of war. Applicants must apply for a commission in the Naval Reserve and accept it, if offered.

Inactive Naval Reserve Commissions are granted to those graduates who meet all qualifications, and graduates who desire to serve on active duty are normally assigned to billets of their choice, commensurate with the needs of the service.

It should be noted that the Navy does not consider the Maritime Academies as a primary source for active duty officers and, in effect, discourages recruitment of Maritime Academy graduates. However, a number of graduates each year elect to go on active duty to pursue a career as commissioned officers in the U.S. Navy.

Naval ROTC

In order to provide wider career choices and possibilities, negotiations are underway to establish an NROTC unit at the Academy campus. If these negotiations are successful, undergraduates who qualify will be afforded the opportunity to take advantage of the benefits of such an NROTC unit. These benefits include Department of Defense tuition subsidy and other financial support.

U.S. Coast Guard

Cadets in their senior year may apply for a commission as an ensign, USCGR, in the U.S. Coast Guard, to become effective upon graduation.

Marine-Related Career Opportunities

The broadened curriculums provide for individual election of sequences of courses in marine sciences including oceanography, ocean engineering, fisheries science, and computers. Technical career opportunities for graduates with a background in these marine sciences have been identified with the National Oceanic and Atmospheric Administration (NOAA), the Sealift Service, the Department of Commerce, and private industry and institutions, including oil, mineral, towing, and fisheries industries.

The primary functions of the Placement Office are to counsel and assist Academy cadets in finding suitable employment after graduation. Career opportunities are available in a variety of maritime and professional seagoing positions aboard American-flag oceangoing merchant ships, towboats, oceanographic research vessels, Great Lakes merchant ships, U.S. Army Engineers' dredges, mineral exploration ships and crew boats, harbor and bay ferryboats, fishing. and lobstering trawlers, yachts, and harbor and coast pilot boats.

Career opportunities also exist for qualified graduates as commissioned officers in the U.S. Navy, U.S. Coast Guard, and NOAA.

Many graduates find gainful employment ashore as marine terminal managers, cargo surveyors, marine products and machinery salesmen, field service engineers, marine safety inspectors, shipbuilding expediters, power plant engineers, and similar postions.

The Massachusetts Maritime Academy Alumni Association maintains close relations with the Placement Office, and many career opportunities are passed on through its efforts to the younger graduates.

The Academy Placement Officer also serves as a training representative in placing cadets aboard merchant and government vessels.

SCHOLASTIC REQUIREMENTS

In general, the scholastic requirements for entrance to the Academy are similar to those specified by other colleges or universities for admission to a program leading to the Bachelor of Science degree.

Specific requirements for admission are:

Graduation from an accredited secondary school (college preparatory course) with fifteen units of credit, or equivalent preparation.

The following courses are required for admission and no waiver of these minimum requirements will be granted:

English	3 units
elementary algebra	1 unit
plane geometry	1 unit
intermediate algebra, college preparatory mathematics	1 unit
physics, chemistry, or other natural science	2 units
acceptable electives	7 units

It is recommended that the seven elective units include additional courses from the following subject areas: humanities (English and foreign language), natural sciences (physics or chemistry preferred), additional college preparatory mathematics, and social sciences.

In certain cases it may be possible for an applicant to remove a subject deficiency by the successful pursuit of a summer program at an accredited secondary or postsecondary institution of his choice, prior to entry into the Academy. The Director of Admissions should be consulted to resolve any question in this area.

Acceptable College Entrance Examination Board Scholastic Aptitude Test Scores (verbal and mathematics).

Meeting the minimum scholastic requirements does not necessarily assure admission. Applicants who meet minimum scholastic requirements are posted on an Order of Merit. Position on this Order of Merit is established by the applicant's composite score. Individual composite scores are based upon the high school record, class standing, CEEB and SAT scores, extracurricular activities, work experience, and motivation. Candidates for appointment are selected in order of merit from the top of the list until all vacancies are filled.

A separate Order of Merit is established for applicants who reside outside the Commonwealth of Massachusetts.

ADVANCED PLACEMENT POLICY

The Massachusetts Maritime Academy is a participating institution in the College Entrance Examination Board (CEEB) Advanced Placement Program. Advanced credit may be offered to those students achieving a grade 3 or higher in the CEEB Advanced Placement Examinations in

American History
English
Spanish
Chemistry

Physics B and C
Calculus AB and BC
For further information consult:
    The Office of the Academic Dean
    Massachusetts Maritime Academy
    Buzzards Bay, Massachusetts 02532.

COLLEGE LEVEL EXAMINATION PROGRAM

The Massachusetts Maritime Academy is now an approved CLEP (College Level Examination Program) testing center, and provides for college course credit via examination: with a passing examination score, a student can receive credit for meaningful experiences and independent study, as opposed to spending time in the classroom covering material in which he is knowledgeable.

CLEP general examinations measure achievement in English composition, humanities, mathematics, natural science, social science, and history. Subject examinations are given in over thirty different fields.

These examinations are administered by the Office of the Academic Dean, which may be contacted for further information.

MINORITY PROGRAM

The Massachusetts Maritime Academy actively participates in recruiting minority students. The Admissions Officers and the Director of the AID (Alternatives for Individual Development) Program make a special effort to inquire about minority college candidates. No individual will be refused admission to the college on the basis of race, color, religion, or national origin.

The Director of AID Program will work with the Affirmative Action Officer to encourage and facilitate the admission of minority students to the Massachusetts Maritime Academy.

OTHER ELIGIBILITY REQUIREMENTS FOR ADMISSION

To qualify for consideration for a federal subsidy, an applicant must be a U.S. citizen, unmarried, and remain unmarried during his or her enrollment at the Academy. He or she must also be physically qualified. Newly appointed freshmen must agree in writing to apply during orientation for a commission as ensign, USNR, and to accept such a commission if offered. This commission does not obligate the graduate to active duty upon graduation.

Nonimmigrant Alien Students

The Academy is authorized under federal law to enroll nonimmigrant alien students.

Transfer Admission

The Academy is a signatory of the Massachusetts transfer compact, and welcomes transfer students who wish to pursue a course of study leading to a career in the maritime industry. The number of credits transfer students receive depends upon the character, quantity, and quality of the work they have done.

Transfer students should note that it is ordinarily not possible to satisfy either the marine engineering or the marine transportation degree requirements in less than three years at the Academy.

Program Requirements for Transfer Students

To be recommended for a B.S. degree, all cadets must demonstrate proficiency in the program major offered by either the Marine Engineering Department or the Nautical Science Department, and accumulate a minimum of 180 days of sea time.

Transfer students should note that because of the requirement of at least 180 days of sea time, it is ordinarily not possible to satisfy degree requirements in less than three years at the Academy.

Graduates from two-year colleges would typically enter the Academy as sophomores.

To be eligible for admission as a sophomore, third class student, a transfer student must have obtained the following credits and a grade of 2.0 in each course accepted for transfer, in addition to meeting physical requirements:

Area		Subject	Courses	Credits
basic sciences		alg/trig	1	3
		calc	1	3
		chem	2	6
	or	physics	2	8
humanities		Engl. comp.	1	3
		Engl. lit.	1	3
social sciences		civilization or equiv.	1	3
additional credits		computers, math or additional phys sci.	1	3
		humanities or soc. sci.	4	6–8
			12	30

Transfer students may be waived Introduction to Marine Engineering (2 credits) or Introduction to Nautical Science (6 credits).

APPOINTMENTS

Appointments to the Academy will be made on a modified "rolling" basis. Applicants who have completed applications for the Massachusetts Maritime Academy and the State College System, and who have their high school transcript and College Board Scholastic Aptitude Test scores (verbal and mathematics) forwarded to the Academy will be placed in a "decision-to-appoint" status. If the applicant's background indicates sufficient achievement, he will immediately be notified of his appointment.

Applicants with acceptable qualifications, whose composite score places them below the Academy quota for appointment will be placed in an "alternate" category. Any class vacancies that occur up to two weeks following the freshmen enrollment date will be filled from the list of alternates.

Upon receiving an appointment to the Academy, the applicant must do one of the following within one month:

Inform the Academy that he accepts the offer of appointment and pay the necessary nonrefundable $100 deposit, which will be credited to his account ($50 against the first quarter tuition fee and $50 against the first quarter dormitory charge); or

Notify the Academy that he declines the appointment.

FINANCIAL INFORMATION

Expenses

Current estimated expenses (subject to change) for each cadet during his enrollment at the Massachusetts Maritime Academy are set forth below. All costs are included except those of an incidental or personal nature. Cadets who have been awarded the federal subsidy to help defray the costs of subsistence, uniforms, and textbooks can currently subtract approximately $600 per year from the indicated total annual expenses. Personal expenses for recreation, transportation, laundry, dry cleaning, and the like are not included. Approximately $600 should be allowed for these expenses.

	Freshman	Sophomore	Junior	Senior
Total yearly estimated expenses, including academic activities, athletic and breakage fees, and uniforms	$3680	$2550	$2560	$2340

Financial Aid Principles

The Academy will strive to assist any cadet whose family and personal resources are insufficient to meet the cost of completing the Academy's educational program. Three categories of financial assistance are available: scholarships (subsidies, grants, gifts); loans (with varying interest rates and conditions of repayment); and employment. Generally, the type of aid awarded depends on each cadet's financial circumstances, academic status, and potential. Most financial aid combines these categories to meet the cadet's specific need.

Financial assistance from the Academy is considered to be supplementary to the family contribution. It is expected, therefore, that both the cadet and his or her parents will explore all areas of financial assistance, including state and local scholarships and loans. Financial assistance is provided on the basis of the financial need of the individual cadet. Financial need is defined as the difference between a family's resources and the total expenses of attending the Academy. The data from the Parents' Confidential Statement (PCS) provide an objective analysis of the family's financial resources. By referring to financial and other relevant information concerning the cadet, the Academy makes a judgment as to his or her need and the type of financial assistance to be awarded within the resources available to the Academy. Information concerning family financial circumstances is, of course, confidential.

How to Apply for Financial Aid

Entering freshmen and transfer cadets who wish to be considered for financial aid must complete and return the Financial Aid Application, which is included with the Letter of Appointment to the Academy. In addition, a Parent's Confidential Statement (PCS) must be submitted to the College Scholarship Service, Box 176, Princeton, New Jersey, designating the Academy as a recipient. Both the application form and the PCS should be submitted prior to June 15 to assure consideration for financial assistance. Applications for financial aid (including PCS) must be renewed each year. Financial aid—other than Maritime Administration cadet subsidies—is not automatically renewed. After reviewing applications and family financial information, the Academy makes awards from the most suitable resources. Further inquiries concerning financial assistance should be addressed to the Financial Aid Officer, Massachusetts Maritime Academy, Buzzards Bay, Massachusetts 02532.

THE ACADEMIC PROGRAM

The Massachusetts Maritime Academy offers two four-year curriculums. One leads to a Bachelor of Science in marine transportation and a federal license as third mate, steam and motor vessels, and the other leads to a Bachelor of Science in marine engineering and a federal license as third assistant engineer, steam and motor vessels. Both programs provide the cadet with a sound foundation in mathematics, physical science, humanities, and social studies, as well as in required professional subjects. The Maritime Administration requirement for each cadet to obtain six months training aboard a training vessel in cruise status is met by a combination of approximately four months training on TS *Bay State* and approximately two months on board a commercial vessel. Courses in naval science qualify graduates to apply for a commission as ensign, USNR.

To the extent possible, choice of curriculum is the decision of the cadet at the end of the fourth class (freshman) year. When, however, staff and facility limitations restrict personal selection, certain cadets will be assigned to a specific curriculum based on academic performance during the freshman year.

The Academy employs a semester system. Two academic semesters, approximately sixteen weeks each, and one sea term make up the eleven-month academic year.

Credit is not given for required physical education, military drill, machine shop, sea laboratory-engineering laboratory, or ship maintenance. A maximum of four semester hour credits is given per semester in each course in nautical science and marine engineering science. Eight semester hour credits are given to each sea term, whether deck, engineering, or a combination of both.

The four-year curriculum of the Academy comprises the following elements:

Core Curriculum	Courses in the humanities, social sciences, mathematics, and physical sciences that embody the essentials of undergraduate general education. These courses are required of all cadets.
Major curriculum	Professional courses in the Departments of Nautical Science and Marine Engineering Science identified respectively with the major degree programs of marine transportation and marine engineering.
Support courses	Additional courses offered by all departments relating to and supporting the two major programs and specific minors. Free electives are also included in this category.
Minor electives	Sequential courses presented in the junior and senior years leading to minors in oceanography, ocean engineering, computer science, fisheries science, and mathematics. Additional minors in management and humanities are proposed.
Naval science curriculum	Courses offered by the Naval Science Department qualifying graduates for a commission as ensign, USNR.

Electives

Each cadet has the opportunity to choose five electives in his or her junior and senior years. They must take four from a sequence of courses comprising a minor. The fifth elective is a free elective in the junior year.

The objective of the minor electives is to present courses of sufficient breadth and depth to qualify graduates for potential vocational opportunities in marine-related fields. The objective of the free elective is to enable each cadet to pursue a course of personal interest in specific disciplines beyond courses offered in the core curriculum of support courses.

These courses comprise the minor electives:

Oceanography
  Physical oceanography
  Chemical oceanography
  Biological oceanography
  Geological oceanography

Ocean Engineering
  Marine Resources
  Marine hydrodynamics
  Marine structures
  Ocean engineering instrumentation

Computer Science
  Numerical methods
  COBOL and data processing
  Digital system design
  Advanced programming

Fisheries Science
  Marine ecology
  Ichthyology
  Fisheries techniques and economics
  Fisheries research and management

Mathematics
  Applied calculus I
  Applied calculus II
  Linear algebra
  Probability and statistics

These courses are offered as free electives:

  Organic chemistry
  Nuclear physics
  Great books
  Literature of the sea
  General psychology
  Political science

ACADEMIC STANDARDS

Courses of Instruction

Attendance at all scheduled classes is mandatory unless, for personal or official reasons, a cadet has been excused from class. Cadets must enroll in all required courses of instruction unless exempted by the Dean for reasons of advanced placement examination, course equivalency completion, transfer, CLEP test, or reversion from a previous class. Exemptions may not exceed thirty percent of a semester's academic program.

Grading System

The Massachusetts Maritime Academy employs an alphabetical grading system for each course and a quality point system for computing averages:

Alphabetical Grade	Per Cent Equivalent	Grade Point
A	90–100	4.0
B	80–89.9	3.0
C	70–79.9	2.0
D	60–69.9	1.0
F (Failure)	below 60	0.0

Quality points are computed for each course as the product of grade points and semester hour credits. Quality point averages are computed for all courses at the end of each term, and posted as term quality point averages. These averages are cumulative for successive semesters, and are posted as cumulative quality point averages. A cadet must achieve a cumulative 2.0 for the four years to graduate, a cadet must complete satisfactorily the minimum requirements of either the marine engineering or marine transportation curriculum with a minimum cumulative quality point average of 2.0, participate in and receive a passing grade in training cruises, comprising a total of at least 180 days, maintain prescribed standards of conduct and aptitude, and apply for a commission in the U.S. Naval Reserve and accept that commission if offered.

A cadet who meets these criteria receives the Bachelor of Science degree and, upon successfully passing the appropriate USCG license examination, receives a federal license of third mate or third assistant engineer. Those who receive the degree and the license may qualify for commissioning as ensign, USNR, Inactive.

CADET LIFE AND RESPONSIBILITIES

The cadet way of life makes the Massachusetts Maritime Academy considerably different from the conventional college or university. The cadet wears a distinctive uniform, lives in Academy residence halls or on board the training ship (during the annual cruise), takes his meals in the cadet mess, maintains a rigid daily schedule, and conforms to strict discipline. This way of life involves a gradual development from follower to leader; from virtually no privileges, to minimum supervision; from responsibility to self only to broad responsibility in the management of the Cadet Battalion.

The standards of conduct in cadet life are necessarily high. Offenses that may be condoned elsewhere are intolerable among cadets. Adherence to the principles of honor, personal integrity, and loyalty has traditionally characterized the professional officer. Every officer must follow these principles if he is to perform his duties and carry out his responsibilities properly. Therefore, the requirements regarding the degree of manliness, honor, and integrity of other

Standards of cadet life are high.

schools or communities have no bearing on those that must prevail at the Academy.

In addition to his regular classwork, the cadet is continually being trained for his future as a deck or engineering officer and leader. He develops character by following the previously specified principles and through the application of self-discipline. He practices leadership through the Cadet Battalion organization, the training ship organization, the instruction of new cadets, and participation in competitive athletics. He applies his professional knowledge and leadership

Never a
dull moment
OR STUDENT
at the
MARITIME
ACADEMY!

training in the annual cruises on board the Academy training ship or other merchant ships.

VARSITY INTERCOLLEGIATE ATHLETICS

Expanding the physical facilities opened a new era in varsity athletics at the Academy, with the acquisition of a completely new sports complex consisting of gymnasium (selected by the Boston Celtics for

ATHLETICS

"We are small but...We are MIGHTY!"

preseason training), swimming pool, bowling alley, rifle range, physical fitness room, tennis courts, baseball field, and football stadium. The Battalion of Cadets is represented by intercollegiate teams in football, soccer, basketball, hockey, tennis, lacrosse, baseball, golf, and wrestling. All cadets are encouraged to participate in at least one varsity sport.

The Academy is a member of the following:

Massachusetts State College Athletic Conference
Eastern Collegiate Athletic Conference

184

National Collegiate Athletic Association
Colonial Intercollegiate Lacrosse Conference
International Soccer Association
National Rifle Association
New England Football Conference
Amateur Bicycle League of America

INTRAMURAL ATHLETIC AND RECREATIONAL PROGRAM

The Academy's program of intramural athletics and recreation provides an opportunity for physical, mental, and social development. Many of the activities provide keen competition, exercise, and carry-over skills that cannot be found in any other part of the physical education program. No activity is compulsory, but it is hoped that all activities will be appealing and satisfying so that participation is desired by the individual.

The program consists of the following: touch football, soccer, golf, tennis, bowling, handball (one wall), paddleball, cross-country, badminton, 50-mile swim club, 100-mile jogging club, bike race, street hockey, ice hockey, basketball, boxing, tug-o-war, foul shooting, track meet, swimming, volleyball, softball, 7-mile foot race, wrestling, rifle, pistol, and sailing.

CADET PROGRAMS AND ACTIVITIES

The Academic program at the Massachusetts Maritime Academy is complemented by voluntary cadet participation in the following nonathletic organizations and activities. In each area of interest, a member of the faculty serves as advisor to provide guidance and continuity to the programs: Band, Cadet publications, Christian Union, Ring Dance, Circle "K" Club, Photography Club, Newman Club, Propeller Club, Sailing Club, Rifle Club, Drama Club, and Bicycle Club.

MEDICAL CARE

The Academy maintains a dispensary on campus for outpatient and limited inpatient treatment of enrolled cadets. A licensed surgeon and a pharmacist staff the "sick bay" during the academic year, both at the Academy facility and at sea on TS *Bay State*.

For more prolonged treatment and diagnosis, cadets are eligible to use the facilities at the U.S. Public Health Service.

ACCIDENT INSURANCE

A group accident plan is provided covering all cadets.

GROUP LIFE INSURANCE

A life insurance program is provided for all cadets.

APPLICATION

Admission to the Academy is based entirely on the qualifications of the applicant, and appointments are made without regard to race, color, creed, national origin , or requirement for financial assistance.

Requests for application forms and information should be directed to the high school guidance office, or the Director of Admissions, Massachusetts Maritime Academy, Buzzards Bay, Massachusetts 02532, preferably no later than the fall of the applicant's senior year in high school. Completed application forms with all required information must be submitted not later than March 1 of the year in which entry is desired.

The Massachusetts State College application is to be completed and forwarded according to instructions provided in the application package. A $10 nonrefundable fee, payable to the Massachusetts State College System must accompany this application.

In addition, a Massachusetts Maritime Academy application form must also be completed and mailed directly to the Academy's Admission Office. Upon receipt of this form the candidate's application will be acknowledged and the following information will be requested:

1. Birth certificate or evidence of citizenship,
2. College Entrance Examination Board Scholastic Aptitude Test scores (verbal and mathematics),
3. Complete transcript of scholastic record from any high school, preparatory school, or college attended by the applicant. If the senior year of high school has not been completed, the transcript must include marks for the first marking period of the senior year.
4. Report of recent chest X-ray, and other medical information including Report of Medical History (form provided), record of inoculations and dates, and prescription of eyeglasses (if worn).
5. Record of arrests and convictions (form provided).

The Maritime College of the State University of New York

The Maritime College is a specialized college of the State University of New York that educates and trains young men and women for employment as licensed officers in the American merchant marine and for professional positions ashore or afloat in the maritime and related industries.

Aerial view of New York Maritime College.

The program at Maritime combines formal academic studies leading to the B.S. or B.E. degree, coursework, and practical experience at sea preparing for the license as third mate or third assistant engineer, and courses in naval science leading to eligibility for a commission as an officer in the Naval Reserve (Inactive) or for active duty in the Navy. The academic studies and the license program also prepare the engineering student for the initial examination leading to registration as a Professional Engineer. All graduates are prepared for graduate technical, business, and law studies.

Almost all Maritime College students are cadets who receive a federal cash subsidy to help offset expenses. Students who are not qualified for the license and subsidy by reason of health or citizenship may enroll at the college as cadets even though they are not entitled to this support. Upon completion of the course they may be certified by the Coast Guard that they meet all requirements for the license with the exception stated.

The unique combination of fully-accredited college majors combined with preparation for the merchant marine license prepares graduates for immediate employment in positions of responsibility afloat or ashore. The technological advances, the complexity of international business operations, and the many unknowns of the hydrospace environment promise a future with challenging careers for young people with the professional competence, academic training, team experience, and self-discipline that can be developed at the Maritime College.

187

BUILDINGS AND FACILITIES

Fort Schuyler

Originally constructed in the early 1830s to protect New York City from attack by water from Long Island Sound, Fort Schuyler has been a maritime school since 1892. It was conveyed to the State of New York by the federal government as a permanent, shore-based home of the Maritime College. The fort is still considered one of the finest examples of Napoleonic military architecture in existence. The interior of the massive stone structure has been modernized and contains classrooms, the library, laboratories, and administrative offices. The inner pentagonal lawn, formed by the walls of the fort, is named in honor of the College's first training ship, the USS *St. Mary's*.

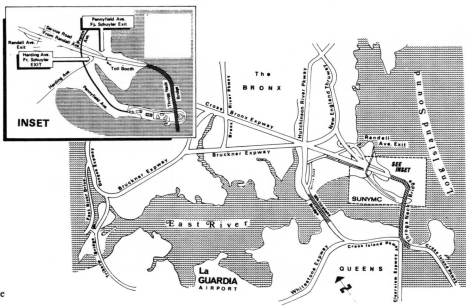

Location of New York Maritime College.

CAMPUS MAP

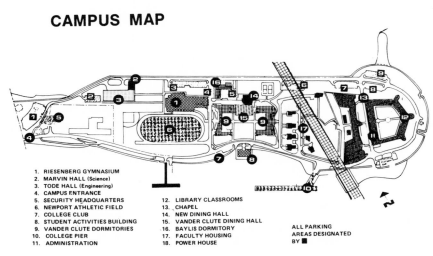

1. RIESENBERG GYMNASIUM
2. MARVIN HALL (Science)
3. TODE HALL (Engineering)
4. CAMPUS ENTRANCE
5. SECURITY HEADQUARTERS
6. NEWPORT ATHLETIC FIELD
7. COLLEGE CLUB
8. STUDENT ACTIVITIES BUILDING
9. VANDER CLUTE DORMITORIES
10. COLLEGE PIER
11. ADMINISTRATION
12. LIBRARY CLASSROOMS
13. CHAPEL
14. NEW DINING HALL
15. VANDER CLUTE DINING HALL
16. BAYLIS DORMITORY
17. FACULTY HOUSING
18. POWER HOUSE

ALL PARKING
AREAS DESIGNATED
BY ■

Empire State

The *Empire State* is the former USNS *Barrett* of the U.S. Navy's Military Sealift Command. She was assigned to the Maritime College in September 1973. The vessel has stateroom accommodations for 500 cadets. The 17,630-ton ship is 533 feet in length, with a beam of 73 feet. Her geared steam turbine develops 13,750 horsepower and a cruising speed of 20 knots.

In addition to subsidy payments to the cadets and the loan of the training ship, the federal government, through the Maritime Administration, makes a direct grant to the college of $75,000 each year, and supplies part of the funds necessary to maintain and repair the training vessel.

Training ship EMPIRE STATE.

HISTORY

The Maritime College celebrated its first centennial on December 10, 1974.

From a modest beginning aboard the sloop of war USS *St. Mary's,* anchored in the East River of New York City, the Maritime College has evolved as an outstanding institution of education for Merchant Marine officers in the United States.

The Governor of New York made prompt application for a ship and the Secretary of the Navy assigned the USS *St. Mary's,* which arrived in New York harbor on December 10, 1874. On that day Captain Luce transferred operational control of the vessel from the United States Navy to the City of New York, and Commander R. L. Phythian, USN, Commanding Officer of the *St. Mary's,* became the first Superintendent of what was then called the New York Nautical School.

On January 11, 1875, the first group of 26 boys boarded her at her berth on East 23rd Street in New York City. By June of that year, she had her maximum complement of 123 aboard.

Candidates for admission had to be between fifteen and nineteen years of age, able to pass the recruit physical of the U.S. Navy, pass a written examination, and have a personal interview. They paid nothing for tuition, quarters, or food. They were required to deposit the sum of $30 to cover costs of uniforms and living expenses.

For the next few years, although the courses in navaigation and seamanship were considerably strengthened, the New York Nautical School remained basically unaltered in character until the turn of the century when, in 1905, courses were added in steam and electrical engineering—another first in the United States.

The Spanish American War, the first conflict in which graduates participated, brought changes to the New York Nautical School. The Regular Navy officers, assigned to the USS *St. Mary's,* were detached and ordered elsewhere by the Navy Department. They were replaced by civilians who were both faculty members and ship's officers, a practice largely followed to this day.

America's overwhelming victory at sea during the war renewed public pride in her seamen. Once again, America revitalized her naval and merchant fleet.

In order to provide training in steam propulsion, a vessel with an auxiliary engine was sought as a replacement for the engineless and aged *St. Mary's.* On October 24, 1907, the USS *NEWPORT,* a barkentine of 1153 tons full load displacement and fitted with a triple expansion engine, was assigned to the Nautical School. Although she was small, her equipment was relatively new as she had been commissioned on October 5, 1897.

The *St. Mary's* was soon decommissioned, and finally scrapped in 1908.

On November 1, 1913, the Nautical School was transferred from the Board of Education of the City of New York to the Department of Education of the State of New York. The new parent agency began casting about for a suitable, permanent shore facility in which to conduct classes when the cadets were not aboard the training ship. In addition, the maritime industry was absorbing graduates at a faster rate than they could be produced, and expanded training space was necessary. The search was to take almost twenty years.

In 1934, the present site at Fort Schuyler on Throgs Neck in the Bronx, of New York City, was obtained. The fort, constructed between 1833 and 1845, was named in honor of Major General Philip John Schuyler, of Revolutionary War fame. Built of Connecticut

Fort Schuyler.

190

granite, in the shape of an irregular pentagon, Fort Schuyler was designed to accommodate a garrison of 1250 men.

Between 1934 and 1938 the old Fort was gutted; thousands of cubic yards of earthen works removed, dormitories, offices, library, and classrooms created within its pentagonal walls, and a pier and powerhouse constructed. When renovated, the fort, which had never fired her guns in anger, was considered the finest physical facility for the training of Merchant Marine officers in the world.

The engineer in charge was to later pattern the world famous Pentagon in Washington, D.C., after Fort Schuyler.

With the acquisition of a shore-based campus, academic changes were also begun. On June 30, 1937, the Superintendent, Captain J. H. Tomb, USN (Ret), proposed the first four-year baccalaureate program for Merchant Marine cadets.

On October 1, 1940, the first three-year class entered. And although scheduled to graduate in 1943 with academic credentials, the December 7, 1941 attack on Pearl Harbor thwarted this plan for the first three-year class. The class graduated one year earlier, in 1942. Captain Tomb was later appointed Superintendent of the United States Merchant Marine Academy, known by many as "Kings Point," when the institution was founded.

When America entered World War II, the academic program was suspended and the curriculum was reduced to an eighteen-months program. The campus became a beehive of activity. Soon, approximately 2000 Naval Reserve officers and 500 Merchant Marine cadets were in training at Fort Schuyler.

Following World War II, under the leadership of Arthur Tode, the Chairman of the Board of Visitors, plans were made to resume the baccalaureate program. In February 1946, the Board of Regents of New York State gave provisional approval for the Bachelor of Marine Science degree and on October 8, 1947, the first such degree was awarded at Fort Schuyler to Eugene Norman Starbecker.

Two years later, in March 1948, the New York State Legislature passed acts to create a corporation within the State Education Department, known as the State University of New York, soon to grow into the largest State University in the world. The Maritime Academy, as the College was then named, became one of the original thirty-two colleges of the University. In this new family the College was to experience considerable academic growth.

At the outset, baccalaureate degrees in marine engineering and marine transportation were awarded, and, in keeping with the specialized nature of the College, professional nautical training continued to be provided, to enable the graduates to pass the federal license examinations required for Merchant Marine officers.

It rapidly became evident that other curriculums were necessary in order to meet the needs of the maritime industry and to satisfy the intellectual curiosity of the student population. The first of the new programs was a degree effort in meteorology and oceanography, soon to be followed by nuclear science, naval architecture, electrical engineering, mathematics, and a humanities concentration.

Marvin-Tode Science and Engineering Building.

In 1968, the Maritime College entered the field of graduate education. A graduate study program, leading to the degree of Master of Science in Transportation Management, was begun. The program stresses independent work and is accessible through evening and weekend classes to those fully employed in the transportation industry. Current enrollment is 114. In the first three commencements for this program, 34 degrees were awarded. This program has been outstandingly successful and is widely acclaimed in the maritime industry. A second and complementary graduate program leading to the Master of Science in Transportation Engineering is now under development.

STUDENT LIFE

The Regiment of Cadets is housed two cadets to a room in Vander Clute and Baylis Halls. Cadets and noncadet members of the student body attend the same classes and participate together in athletics and other activities.

All students are expected to be well-dressed and well-groomed while on the college grounds or at college activities. All students are also expected to maintain dormitory rooms in a clean and neat condition.

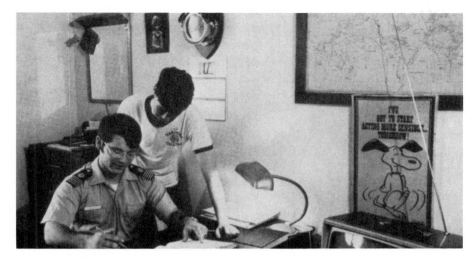

First Class Cadet room.

THE REGIMENTAL SYSTEM

The routine of the daily activities of the Maritime College is conducted on a Regimental basis. Cadets are preparing themselves for licensing as officers in the Merchant Marine and commissioning in the U.S. Naval Reserve, Inactive Duty. The Regimental system is designed to develop leadership and instill in the Cadet an abiding sense of honor, discipline, responsibility, and mature citizenship. Cadets are required to accept responsibility both during the academic year and during the summer sea training period.

Cadet Indoctrination Week

Prior to the commencement of the fall semester, incoming cadets undergo a period of intensive indoctrination into the life at the College within the Regiment of Cadets. Along with receiving an introduction into the College routine, cadets are trained in fundamentals of military organization, including formations, customs, courtesies, and the manual of arms. It is during this period that cadets are furnished with uniforms and the appropriate college grooming standards.

The new cadet is introduced to the administration and various faculty members at the College, and is given a full tour of campus facilities. During this period cadets learn part of the history of the Maritime College and Fort Schuyler, and receive lectures and practical experience aboard the TS *Empire State*. This is a full and rigorous period as it acquaints the cadets with regimental life and prepares the cadets for the academic year.

The Cadet's Day

During the academic year a cadet's normal weekday begins at morning formation at 7:30 A.M. for muster, inspection, and the observing of morning colors. Classes are conducted from 8:00 A.M. to 3:40 P.M., with an intervening lunch period. From then until dinner at

Introduction into the college routine.

193

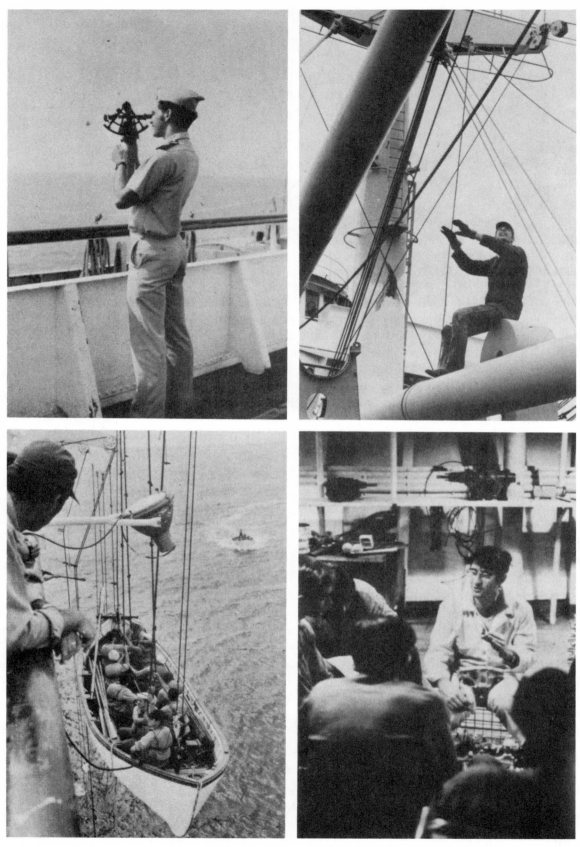

Lectures and practical experience.

5:30 P.M. to 6:30 P.M., the cadet is free to participate in the college intercollegiate athletic program, the intramural program, or any of the extracurricular activities.

Uniforms

Cadets wear uniforms at all times except when leaving on, or returning from, liberty or vacation periods. The tropical khaki uniform is normally worn during spring, summer, and early autumn; navy blue uniforms are prescribed for the late autumn and winter months. Applicants who have been accepted at the College are required to come to the College during the last two weeks of June for uniform fitting so that uniforms may be issued during the indoctrination period.

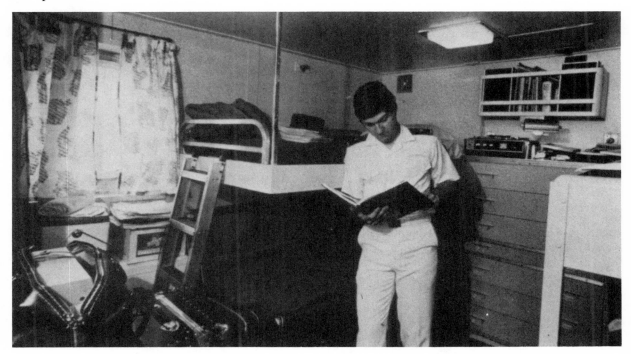

First Class Cadet stateroom aboard T.S. EMPIRE STATE.

Leaves and Special Privileges

The academic year begins on the day after commencement in May and continues to the next graduation day. The months of May, June, and July constitute the summer sea term for all classes. Fall and spring semesters of approximately fifteen weeks duration complete the academic year, with leaves at Thanksgiving, Christmas (semester break), spring break, and after the summer sea term.

The Maritime College assumes no responsibility for dependents of cadets who are married, nor does a married cadet receive any special privilege or advantage not accorded an unmarried cadet.

During the academic year, liberty is normally granted each evening to first class cadets, and on Wednesday evenings to second and third class cadets. All cadets are normally granted weekend liberty from Friday afternoons until 7:30 A.M. the following Monday. Special liberty may be granted for emergencies and personal reasons. Granting special liberties will be based upon justification and the record of the cadet requesting such liberty. Requests for a leave of absence for special reasons must be addressed to the President of the College, and submitted via the Dean of Students.

HEALTH CARE

The college maintains a well-equipped dispensary with a trained medical technician in charge and a physician in attendance during part of each academic day. Primarily the dispensary provides free first-aid treatment and medications for nonserious physical ailments.

Maritime Service Cadets

Cadet students who are enrolled as cadets in the Federal Maritime Service receive medical services and hospitalization through the U.S. Public Health Service.

CHURCHES and SYNAGOGUES

Churches of most religious faiths and synagogues are located in the area adjacent to the college. Special provisions are always made to allow cadets to participate in the religious services of their faiths.

STUDENT ORGANIZATIONS

Extracurricular activities play an important part in the college life of

"Pass in review."

students. Through these activities students obtain experience in the organization and administration of activities that include a variety of interests. Students are encouraged to participate in at least one extracurricular activity in order that they may experience the benefit of group action.

These activities include Student Council, College Band, *Porthole* (college newspaper), *Eight Bells* (college yearbook), Newman Club, Protestant Club, Jewish Society, French Club, Spanish Club, Propeller Club, Naval Architects and Marine Engineers Club, Glee Club, Company P-8, Pershing Rifles, International Relations Club, Student American Nuclear Society, Sailing Squadron, Radio Club, Circle K Club, Science and Engineering Club, Steamship Historical Society of America, Dance Committee, Computer Club, Ocean Science and Technology Society, Judo Club, Eagle Scout Fraternity, Skin and Scuba Diving Club, and Humanities Club.

ATHLETIC PROGRAM

Athletic competition is an integral part of many students' lives. At the Maritime College students have the opportunity to participate in both intramural and intercollegiate competition in a wide variety of sports. While the objective of all athletic programs is team work and friendly competition, many Maritime College students have distinguished themselves, earning recognition as All-American athletes in sailing, swimming, and wrestling.

McMurray Hall, student activities, building and boathouse.

As a small college with an unusually large number of intercollegiate teams, the Maritime College has few bench warmers. Students who desire to compete have the opportunity to be full members of the team.

The Maritime College is a keen competitor in intercollegiate sailing as well as yacht racing. College boats include fleets of modified 420 type dinghies, 30-foot Shields keel sloops, and the 42-foot cruising class sloop *Resistance*.

The college is represented by varsity intercollegiate teams in basketball, cross-country, soccer, sailing, crew, rifle, fencing, tennis, swimming, wrestling, lacrosse, golf, and baseball.

The intramural sports program includes basketball, wrestling, flag football, volleyball, softball, swimming, handball, squash, and paddleball. Tournaments are held and champions are named in each event.

In addition, the Maritime College has a Judo, Ju Jitsu, and Karate Club, which competes informally with other colleges.

The official colors of the Maritime College are navy blue and cardinal red, with white used for trim and contrast.

First woman graduate, Class 1974.

ADMISSIONS

Admission to the Maritime College is based solely on the qualifications of the applicant, and is granted without regard to race, color, creed, sex, or national origin. Successful applicants must meet the requirements for admission as stated below. As admission is competitive, decisions are based on quality or strength of preparation, performance in high school or college, test scores, and other indications of aptitude revealed in activities, hobbies, and experiences.

SCHOLASTIC REQUIREMENTS

Applicants must be high school graduates or holders of a high school equivalency diploma, and present at least sixteen units of credit, unless state or local requirements for graduation differ.

The following courses are the minimum required for admission without exception:

English	4 units
Elementary algebra	1 unit
Plane geometry (or equivalent)	1 unit
Plane trigonometry—½ unit (or equivalent)	1 unit
Intermediate algebra—½ unit	
Physics or chemistry	1 unit

Students are encouraged to pursue mathematics and science beyond the required minimum. The remaining required units can be in social studies, science, mathematics, and foreign languages. Units in other subjects will be individually evaluated. Mechanical or technical drawing is suggested, particularly for engineering applicants.

One of the seven women cadets, Class 1978.

"Hands on experience" in *Empire State*.

ENTRANCE EXAMINATIONS

Applicants who are attending school in New York State will normally take the Regents Scholarship and College Qualifying Test at their high school in October of their senior year.

Scores from at least one of these examinations are required as part of the application: the Regents Scholarship Exam (RSE), or the Scholastic Aptitude Test (SAT) of the College Entrance Examination Board, or the American College Testing Program (ACT) are acceptable.

All applicants are requested to submit SAT or ACT scores taken during the junior or senior year of high school.

Cadet (Not Physically Qualified)

Applicants who do not meet the physical requirements for license as an officer in the Merchant Marine may attend the Maritime College, as a nonsubsidized cadet. A student in this category will take degree and license courses and may be certified by the Coast Guard that he or she meets all requirements for licensure but physical. Such graduates may request a waiver of physical requirements from the Coast Guard. Since physical waivers for license are determined on an individual basis at the time of applying for the license, cadets who are not physically qualified cannot be assured of receiving a waiver. Although not physically qualified for licensure in the Merchant Marine, cadets may be eligible for commission in the naval Reserve because physical requirements for the two programs differ, especially with regard to vision.

Transfer Students

The Maritime College welcomes students who wish to transfer from other colleges. Curriculum structure and federal requirements for license preparation are such that transfer students cannot ordinarily complete the degree and license requirements in less than three years. The third year, even for transfers with more than one year of college experience, is difficult due to the special nature of the Maritime College education as embodied in the combination of degree and license preparation.

Transfer credit will be given to the maximum possible, as appropriate to the Maritime curriculum. The Admissions Office would be pleased to assist prospective transfer students plan their preparation.

Graduates of two-year programs in engineering science and technology and graduates of similar technically-oriented programs are particularly well suited for transfer.

Veterans

In recognition of the unique experiences and special needs of veterans, the Maritime College invites veterans to inquire about special programs leading to the degree and the merchant marine license. Inquiries can be made to the Director of Admissions.

Day Students

To be considered for admission as a day student, a student must be at least twenty-five years old, and must have either a valid federal merchant marine license issued by the Coast Guard or have served enough time at sea to satisfy the USCG requirements to sit for the license. An exception may be made if it is possible for a student to complete the USCG requirements by sailing during the summers while attending the Maritime College. This sailing experience cannot be acquired aboard the College's training ship.

Day students may reside on campus. Day students are not eligible to receive the federal subsidy.

Early Admission Program

The Maritime College has been authorized by the New York State Education Department to admit students who have completed their junior year in high school, provided they meet the guidelines cited below. Because this is a big step for a high school junior, interested and qualified students are encouraged to arrange for an interview to explore the appropriateness of such a move. Students accepted to this program should make arrangements with their high school so they can receive their high school diplomas. For consideration, students must:

1. Have completed the minimum requirements in mathematics and science;
2. Be recommended by the high school counselor and a teacher;
3. Have a minimum high school average of 85 percent in the sophomore and junior years;
4. Have minimum SAT scores of 550 verbal and 600 math; and
5. Be deemed sufficiently mature, by the college, to meet the demands of college life.

Foreign Students

The Maritime College welcomes students from other countries. In 1975–76, 151 students from 16 foreign countries were in attendance. Foreign students are enrolled as cadets pursuing degree and license courses. At graduation, foreign students receive the B.S. or B.E. degree and may be certified by the Coast Guard that they meet all requirements for license except citizenship.

Advanced Standing

Advanced standing may be granted to students who participate in the various programs cited below. Students who have a strong background in a subject might consider these programs as a means of permitting enrichment of their studies while at Maritime. In addition to the advanced placement programs listed below, students may achieve advanced standing or advanced placement on the basis of appropriate examinations and other means of evaluation established by a department. Students who are granted such standing will ordinarily be exempted from a corresponding number of credits for graduation.

Advanced Placement Program (CEEB) Advanced placement may be extended to new freshman students who have completed courses in secondary school in the Advanced Placement Program of the College Entrance Examination Board. Candidates undertaking Advanced Placement courses in secondary school are expected to take the appropriate examinations and to request that their scores be forwarded to this institution.

College Level or Proficiency Examinations It is the policy of the Maritime College to grant credit for satisfactory performance in the College Level Examination Program of the College Entrance Examination Board and the College Proficiency Examination Program of the New York State Education Department whenever such examinations cover material given in a similar course at the Maritime College. Credits will be granted by the Office of the Vice President for Academic Affairs upon the recommendation of the chairman of the department in which the course is taught.

Service-Connected College Level Experience The College grants credits on a selective basis for formal service school courses listed in the publication "A Guide to the Evaluation of Educational Experiences in the Armed Services." Credit may also be given for United States Armed Forces Institute (USAFI) courses. The General Educational Development (GED) test scores may be accepted for admission of non high school graduates.

Summer Institute

A Summer Institute is available for four weeks during late July and early August, and is recommended for incoming students who need or desire additional work to better prepare for their collegiate studies. The Institute is designed to provide instruction in English, mathematics, and study skills, in order to make easier the transition from high school to college. Instruction is scheduled during the mornings, five days a week. Students may reside in the college dormitories. Additional information can be obtained from either the Admissions Office or the Office of Continuing Education.

EXPENSES

Tuition and Fees

The following is a schedule of estimated expenses for full-time undergraduate students (tuition for part-time undergraduate and graduate students is also included) for the academic year 1977–78.

A. Tuition

I. Full-time* undergraduate students pursuing complete cadet program (except foreign students†):
1. Lower Division** (per semester) $375.00

2. Upper Division (per semester)		450.00

II. Full-time undergraduate students *not* pursuing cadet program:

1. New York State Residents		
Lower Division (per semester)		$375.00
Upper Division (per semester)		450.00
2. Out-of-State Residents		
Lower Division (per semester)		600.00
Upper Division (per semester)		750.00

III. Part-time undergraduate students:

1. New York State Residents	
Lower Division (per credit hour)	$ 25.00
Upper Division (per credit hour)	30.00
2. Out-of-State Residents	
Lower Division (per credit hour)	40.00
Upper Division (per credit hour)	50.00

IV. Graduate Students: Full-time:

1. New York State Residents	$700.00
Out-of-State Residents	900.00

V. Graduate Students: Part-time:

1. New York State Residents (per credit hour) $ 58.50

2. Out-of-State Residents (per credit hour)   75.00

B. *College fee* (per semester)	$ 12.50
C. *Student activity fee* (per semester)	35.00
D. *Room in residence hall* (per semester)	375.00
E. *Board* (per semester)	450.00
F. *Summer cruise charges* (including laundry fee)	362.50
G. *Service charge for late registration*	10.00
H. *Service charge for late payment of fees*	20.00
I. *Returned check charge*	5.00
J. *Transcripts*	2.00
K. *Key deposit* (fall semester semester only)	2.00
L. *Damage deposit* (first semester only)	50.00

Other Costs

Uniforms will normally cost about $1000 for the four years, $800 is payable during the summer before the beginning of the freshman year, and $60 is due about April 1 of the freshman year. The balance would be for replacements during the four years.

* Full-time student—enrolled for twelve or more credits per semester.
** Lower Division—must have completed fifty-six or fewer semester hours.
·† Foreign students must pay out-of-state tuition.

It is necessary to budget about $700 per year for personal expenses, clothing, travel, and recreation, and an additional $150 per year for books and educational supplies. These are approximate costs subject to increase.

Financial Assistance

The college endeavors to assist as many students as possible who do not have sufficient resources to meet all college expenses. The amount of financial aid awarded to a student is determined by analysis of actual financial need.

It is expected that parents will contribute as much as possible in meeting expenses before requesting financial assistance. This contribution is judged on the basis of the Parent's Confidential Statement (PCS), explained later in this section.

Applicants for financial assistance are accepted and awards made without regard to race, color, creed, sex, or national origin.

Parents' Confidential Statement (PCS)

The college employs the College Scholarship Service in considering applications for all scholarships and loans. Therefore, any candidate for a loan, grant, or scholarship, except New York State Tuition Assistance program, Guaranteed Student Loans, and the federal subsidy, must submit a Parents' Confidential Statement (PCS) form to the College Scholarship Service (CSS) designating the Maritime College as a recipient.

The Parents' Confidential Statement should be submitted to CSS as early as possible but prior to May 1. Forms for entering students are available at high schools, or by writing to College Scholarship Service, P.O. Box 176, Princeton, N.J. 08540. Forms for students in attendance are available at the Financial Aids Office.

Federal Subsidy

The Maritime Administration provides federal subsidy payments at the rate of approximately $50 a month to each eligible cadet while in attendance. Payment is made in three installments during the year, usually about the first day of September, December, and May. In the case of the fourth class (freshman) year, the first payment occurs in December.

To be eligible, each cadet must be a U.S. citizen and comply with the age and physical requirements outlined in admission procedures. Each cadet must also agree in writing to apply for and accept an inactive commission as an ensign in the U.S. Naval Reserve. He must also enroll in and complete the naval science courses.

ACADEMIC REGULATIONS

Admiral's List and Dean's List

A student with a semester average from 4.0 to 3.5 is eligible for the Admiral's List. A student with an average from 3.4 to 3.0 is eligible for the Dean's List. A minimum of 15 credits must be carried for such recognition.

Grading

Grade		Quality Points
A	Excellent	4
B	Good	3
C	Satisfactory	2
D	Passed	1

A 2.0 average is required for graduation. Students with semester or cumulative averages below 2.0 will be reviewed by the Academic Board. Students are subject to academic disenrollment if their cumulative average is below standard. The standard for disenrollment is increasingly rigorous as progress is made toward graduation. Typically, the Board will require semester averages of 1.4 and 1.7 in the freshman year, 1.8 for the first sophomore semester, and 2.0 thereafter.

Registration

Students must register in person during the preregistration and prior to the first day of classes each semester. A cadet cannot be credited as qualifying for the federal subsidy until his registration card is deposited with the Registrar.

Transfer Credits

Credit for courses taken at other colleges may be transferred if the course is comparable in level and content to the Maritime College curriculum, and if a grade of at least C was earned. Evaluation of transfer credits is made upon receipt of a final official transcript and a catalog that includes the course description. Students must make certain these documents are sent to the Registrar of the Maritime College. Grades and honor points are transferred for Maritime students completing extra work at another college.

Graduation

A cumulative average of 2.0 is required for graduation. All requirements must be completed before a diploma is awarded. It is the responsibility of the student to insure that he or she has satisfied all requirements by the time of graduation. Those who do not graduate in May will be awarded their degrees on the graduation dates in the academic calendar. Pending the receipt of this diploma, a student who has completed all requirements will be issued a letter attesting to having met all requisites for graduation, if such is necessary for employment. All tuition fees, etc., must be paid prior to release of the diploma or transcripts.

UNDERGRADUATE ACADEMIC PROGRAMS

License programs lead to the degree of Bachelor of Science or Bachelor of Engineering, the U.S. Coast Guard-issued Merchant Marine license as third mate or third assistant engineer, and eligibility for commission as ensign in the U.S. Naval Reserve, Inactive. Students who are not physically qualified for the license will earn the degree and may receive certification by the Coast Guard that they meet all license requirements but physical.

Nonlicense programs lead to the degree of Bachelor of Science or Bachelor of Engineering. Admission is limited.

Day students, since they must hold a Merchant Marine license, will follow modified curriculums generally omitting license preparation subjects. Credit hours required for graduation are adjusted accordingly. Students should consult their degree-granting department and the registrar.

Engineering Department

Engineering is defined as the profession in which knowledge of the mathematical and natural sciences, gained by study, experience, and practice is applied with judgment to develop ways to utilize economically the materials and forces of nature. The engineering curriculums offered at the Maritime College are designed to supplement a sound theoretical education with practical background, by extensive use of laboratories both afloat and ashore. In addition to learning the principles of this challenging and satisfying profession, the student is given an appreciation of past and present society, together with training in effective communication.

The Bachelor of Engineering degree is offered in electrical engineering, marine engineering, naval architecture, and ocean engineering.

Marine Transportation Department

The course of study for cadets majoring in marine transportation includes the theoretical and practical education necessary to develop highly qualified licensed officers. It combines the humanities and sciences with nautical and marine transportation subjects to achieve a well-rounded collegiate program that will fully equip the graduate to

Bridge watch.

meet the present and future problems of life and the needs of the maritime industry, afloat and ashore.

The marine transportation curriculum offers two concentrations, transportation economics and transportation management.

Science Department

The Science Department offers curriculums in the following:
    Nuclear science (engineering license)
    Meteorology and oceanography (deck license)
    Computer science-mathematics (engineering license)
    Computer science-mathematics (deck license).

"At the wheel".

Nuclear Science All estimates indicate that a large number of nuclear power plants will be built in the U.S. and abroad in the next few decades. These power plants will be used to produce electfical energy, to power ships, to desalinate water, and for other purposes. The nuclear science curriculum includes extensive preparation in mathematics, physics, chemistry, and metallurgy, as well as in the humanities and social sciences. The complete set of courses is designed to establish a firm foundation for a professional career in the immediate and foreseeable future. The individual courses have been designed to prepare cadets for the professional duties involved in operations and research and development in nuclear energy, as well as to prepare them in the necessary practical nautical training required to become licensed ship officers. The courses in reactor theory, together with the basic engineering courses and the practical training with the ship's engine room, provide a solid and broad base for a career in nuclear power.

Meteorology and Oceanography The meteorology and oceanography concentration is available to cadets who are candidates for the

Bachelor of Science degree and the federal license as third mate in the Merchant Marine. It is designed to provide the deck officer with the academic background in marine meteorology and physical oceanography useful for professional employment in these or related fields, or as a mate/scientist aboard an oceanographic vessel, or for general use in the normal course of sailing with the Merchant Marine.

Computer Science-Mathematics (Engineering License) Computer science is one of the fastest growing areas of learning in our society. Industrial automation (including ship automation) and scientific research are but two of the many examples of the use of this science. The computer science-mathematics curriculum includes several semesters of computer science along with courses in basic and applied mathematics.

The sequence of courses in computer science and mathematics, together with the courses in physics, chemistry, and engineering, and courses in the humanities and social sciences, is designed to prepare the cadet for professional careers in computer science and allied fields of mathematics, in addition to sailing as an engineer in the Merchant Marine.

Application

Prospective students who are attending a New York State high school should obtain a State University of New York application form from their high school guidance counselor, and request supplemental application materials by writing or calling the Admissions Office, State University Maritime College, Fort Schuyler, Bronx, New York 10465. Applicants who are attending school elsewhere, or who have graduated from high school may obtain application forms by writing to the Admissions Office. Students may begin work at the Maritime College only in the fall semester.

Computer Science-Mathematics (Deck License) A computer science-mathematics curriculum with the same courses in computer science and mathematics as for engineers is also available to cadets pursuing training as licensed deck officers.

The courses of study include the practical and theoretical nautical training prerequisite for such officers, together with a series of courses in computer science-mathematics designed to prepare the cadet for a professional career in computer science and in allied fields of mathematics, in addition to sailing as a deck officer in the Merchant Marine. Electronic computers play an established role in the management and operations of the maritime industry, and the use of computers is growing.

Summer Sea Term

The summer voyages are required, credit-bearing, staff supervised educational periods intended to insure necessary operational experience for all cadets. In order to graduate, every cadet must successfully pass each of the three indicated summer periods on the college training vessel, and must have a weighted cumulative average (WCA) of at least 70.0 for the three summer terms. The sea period grade is a report for a single course that is made up of several phases and requirements. Failure may result in repeating the period or disenrollment for inaptitude. Each succeeding period at sea will demand of the cadet the assumption of additional supervisory responsibilities and advanced operational knowledge.

Graduation Day.

The California Maritime Academy

The California Maritime Academy is a unique educational institution. Whereas most schools of higher education emphasize academia, Cal Maritime attempts to strike a harmonious balance between academic, theoretical learning and the practical application of that learning.

As a result, Cal Maritime students are highly sought after by the maritime industries they serve. Although the employment opportunities of the graduates are subject to the whims of the economy, Cal Maritime is extremely proud of the employment record for its classes, particularly the ones of the last five years. Probably few educational institutions in California, and possibly in the United States, can match the employment prospects enjoyed by their graduates.

In this technologically-changing world, those students who have had the advantage of a fully rounded technical education are better prepared to take full advantage of career opportunities.

The advent of fast, highly sophisticated vessels requires a new breed of officer—one who has the breadth of education and depth of training to cope with the complexities of a rapidly changing technology.

California Maritime Academy affords the student not only educational and training opportunities, but also the opportunity to develop into a mature individual, capable of assuming the great responsibility and leadership required by a highly sophisticated Merchant Marine.

The Mission of the California Maritime Academy is to provide instruction in nautical industrial technology, marine engineering technology, and related fields, including all of those necessary to provide the highest quality licensed officer for the American Merchant Marine and California maritime industries.

HISTORY

The California Maritime Academy was established in 1929 as the California Nautical School by an act of the State Legislature. In 1972 it was given its present status as an independent institution of higher education, deriving certain administrative support from the Trustees of the California State University and Colleges.

Federal authority and encouragement for state maritime academies date from an Act of Congress of 1874. While it is distinctly an educational agency of the State of California, the California Maritime Academy obtains considerable assistance from several federal agencies: Maritime Administration, Navy, Coast Guard, and Public Health Service.

The United States Maritime Administration's interest stems directly from a mandate of the Congress, expressed in the Merchant Marine Act of 1936, which directs the maintenance of an adequate Merchant Marine to support American domestic and foreign commerce, and to meet the requirements for national defense. The act provides that the Merchant Marine be "manned with a trained and efficient citizen personnel.''

LOCATION

The California Maritime Academy is located on the north shore of the Carquinez Strait, in the City of Vallejo. It is about a thirty-minute drive on U.S. Interstate Highway 80 from San Francisco. The Naval Shipyard at Mare Island is in the immediate vicinity, and is available for observation of drydocking, heavy shop practice, ship repair procedures, and electronic developments. Oceangoing steamers from all parts of the world pass through the Carquinez Strait en route to and from Sacramento and San Joaquin River ports.

Aerial View of California Maritime Academy waterfront.

FACILITIES

The Academy is situated on a sixty-seven-acre campus adjacent to the Carquinez Strait. A deep-water pier provides berthing space for the training ship *Golden Bear,* and encloses a boat basin for power, sailing, and rowing boats.

There is a well-equipped gymnasium weightroom and twenty-five-meter indoor pool.

Tennis and handball courts and an athletic field provide ample outdoor recreation facilities.

The Academy's training ship *Golden Bear* is a 7987 gross-ton vessel that can cruise at sixteen knots and serves as a ''floating laboratory'' during the annual ten-week training cruise.

Students normally enter the Academy in the fall trimester at the fourth class (freshman) level. Students who have attended a two-year or four-year college, and have taken appropriate courses may enter in the fall trimester at the third class level. Students with lesser or greater amounts of transfer credit should contact the Academic Dean

"Coming into port."

211

at the Academy to determine their entry status and appropriate time of entry.

GENERAL QUALIFICATIONS FOR ADMISSION

Age

Candidates for freshman (fourth class) standing must be at least seventeen and under twenty-four if nonveterans, or under twenty-seven if veterans at the time of entrance into the Academy. Transfer applicants, who are eligible to enter at the second year (third class) level, may be one year older.

Citizenship

All candidates who expect to obtain a Coast Guard license are required to be citizens of the United States. The California Maritime Academy observes scrupulously the requirements of Title VI of the Civil Rights Act of 1964. Section 601 of this title stipulates: "No person in the United States shall, on the ground of sex, race, color, creed, or national origin, be excluded from participation in, be denied the benefits of, or be subjected to discrimination under any program or activity receiving federal financial assistance." Eligibility is without restriction as to sex, race, color, creed, or national origin.

Physical Requirements

Candidates must meet the physical requirements for licensed officers in the U.S. Merchant Marine. Applicable regulations include the following:

1. Eyesight—nautical industrial technology majors must have at least 20/100 in each eye, correctable to 20/20 in one eye and to at least 20/40 in the other. Marine engineering technology majors must have at least 20/100 in each eye, correctable to at least 20/30 in one eye and 20/50 in the other.
2. General health—candidates must be mentally and physically sound. Epilepsy, insanity, badly impaired hearing, or any other disability that might prevent the candidate from performing the ordinary duties of an officer at sea would preclude admission.
3. Color blindness—both nautical industrial technology and marine engineering technology students must be able to distinguish red, blue, green, and yellow in order to apply for the appropriate license.

Naval Reserve

The U.S. Maritime Administration requires that candidates must agree in writing to apply before graduation for a commission as ensign in the U.S. Naval Reserve and to accept the commission if offered.

Application for USCG Documentation

All students will be required to apply for a U.S. Merchant Mariner's Document, which is issued by the U.S. Coast Guard. Additionally, all graduates will be required to apply for a license issued by the Coast Guard. In applying for said document and license, each person must certify that he or she has not been convicted by any court (including a military court) for other than a minor traffic violation, and that he or she has neither used narcotics, nor been addicted to the use of narcotics. The definition of narcotics includes marijuana. A false application in this regard is a federal crime, and any license or document falsely obtained from the Coast Guard may be administratively revoked by that agency.

Scholastic Requirements

Applicants for entrance at the fourth class (freshman) level must be high school graduates or holders of a high school equivalency certificate.

Each candidate must have his school submit detailed records to the Academy of all completed high school, preparatory school, and college work. To insure timely evaluation of the candidate's qualifications, these academic records should be received by April 30 of the year of desired admission to the Maritime Academy.

In addition, candidates must take one of the three entrance examinations listed below and have the results sent to the Academy. Transfer students who took one of these tests while in high school may have the results sent to the Academy. Transfer students who have less than one full year of college work and who have not taken the tests should take them and have the results sent to the Academy. Candidates having one or more years of college work to establish their performance in college-level studies need not submit entrance test results.

1. College Entrance Examination Board Scholastic Aptitude Test (SAT). For dates and locations where these tests are given, consult your school counselor or contact:

College Board ATP
Box 1025
Berkeley, California 94701
(415) 849-9050

2. American College Testing Program (ACT). This series includes tests in English, mathematics, social studies, and natural sciences. For dates and locations where these tests are given, consult your

school counselor or contact:

The American College Testing Program
P. O. Box 168
Iowa City, Iowa 52240
(319) 356-3711

3. California Maritime Academy Entrance Examination. The Academy offers its own entrance examinations in English and mathematics for those who are not able to take the SAT or the ACT. The examinations are held on the first Saturday of every month from January through June. In addition, if some persons are unable to travel to the Academy for the examination, arrangements can be made through a school counselor to administer the examinations locally. Please contact the Academic Dean of the California Maritime Academy to make arrangements.

When an applicant's file is complete, it is reviewed to determine admissibility. Although all of the data in the file are considered, some items are of paramount importance. For example, the successful completion of two units (two years) of high school algebra is so important as a base for the academic programs of the Academy that an applicant without this qualification probably will not be admitted. Also, good performance in the mathematics and science courses taken and good performance on the mathematical and quantitative portions of the entrance examinations are good predictors of success at the Academy, and are considered very important in the determination of admissibility. A desirable distribution of subject matter in high school would be: three units of college preparatory mathematics, two units of laboratory science, three units of English, two units of history, two units of foreign language, one unit of literature or social science, and two units of electives.

It is appropriate for a student to begin a maritime education at a college near home and then transfer to California Maritime Academy to complete the work. Living at home and attending a nearby college reduces the expense of education and gives the student an opportunity to test his or her abilities and preferences in college-level education. However, the format of education at California Maritime Academy is distinctly different from that at other four-year colleges and it is very important to do careful planning for the transfer.

Four commonly-occurring transfer student situations are described below in an attempt to clarify transfer possibilities.

1. A student attends a community college briefly to rectify deficiencies and then enters the Academy in a fall trimester at the fourth class (freshman) level. At the Academy this student takes the basic fourth class program but with modifications occasioned by some required fourth class courses completed at the community college. Courses that can be taken at a community college and serve this purpose may be selected from the list of required courses. Having these courses completed before beginning the fourth class makes it easier to take courses toward one of the options.

2. A student attends a community college or a four-year college for one and a half to two years and takes the courses required. Then the transfer is made to the Academy in a fall trimester at the third class (sophomore) level.
3. A student attends college for two, three, or even nearly four years searching for a field of interest or pursuing one field and becoming discouraged with employment possibilities in it. By this time he has taken most of the required courses and, settling upon the objective of a maritime education, can enter the Academy in the fall trimester at the third class (sophomore) level.
4. A student attends one of the other maritime academies for one, two, or three years and then decides to attend the California Maritime Academy. Such a student can usually complete his or her education with little loss of time, because the curriculums at the various maritime academies are very similar. One year of residence at the California Maritime Academy and an average of 2.0 for transfer credit is required of such a student who wishes to receive a degree from the Academy.

The recommended time for transfer into the California Maritime Academy is the fall trimester of the third class (sophomore) year.

It is well to remember that the curriculums of the Academy require four years, and that school is in session three trimesters (eleven months) each year. Included in the twelve trimesters are the three cruise, or sea-training, trimesters required by the Coast Guard as qualification to sit for the U.S. Coast Guard license examinations. Because there is only one cruise each year, the transfer student must be in residence at the Academy for three years in order to participate in the three required sea-training cruises. The student transferring at the beginning of the third class year must present academic credit equivalent to the fall and spring trimesters of the fourth class year and the winter trimester of the third class year. This enables the transfer student to go on cruise in the winter trimester of the third class year instead of staying in the classroom on campus as the regular third class students do.

If a student takes the required courses in another college and transfers to the California Maritime Academy in the fall trimester of the third class year, he or she will find courses scheduled so that time conflicts do not occur between the various required courses, and the curriculum requirements may be completed within three years. If a student presents an array of courses that do not include the listed courses, a time schedule disaster will result. This is because so many of the courses at the Academy are sequential, most courses are offered only once each year, and the time schedule is very tight because of the many hours devoted to laboratory work.

The Role of State Legislators

In years past legislators nominated individuals as candidates for admission to the Academy. As a result of Concurrent Resolution No. 64, legislators no longer nominate candidates for admission. Still the Board of Governors wishes to keep legislators involved, and notifies them of successful candidates from their districts so that they will have the opportunity to send letters of congratulations. The Registrar at the Academy will send all required enrollment forms and formal notification of admission direct to the candidate.

COST OF ATTENDANCE

Payment

Total assessed fees, as shown in the schedule below, are due on or before the first day of each trimester. There is no exception to this requirement; according to State regulation a student is not enrolled or entitled to attend classes or receive other services until all fees have been paid. Charges are subject to change without notice. Financial assistance is available, but the student must arrange for financial assistance prior to registration.

Maritime Administration Subsidy

Most students will receive a subsidy of $600 per year from the U.S. Maritime Administration (MARAD). However, the selection of subsidized candidates is not made until after the first trimester, based on fall trimester grades. Those who receive the subsidy will be paid directly in quarterly installments.

Schedule of Annual Fee Payments

	first trimester	*second trimester*	*third trimester*
Tuition*	$135	135	135
Athletic fee	10	10	10
Room	175	175	175
Board	400	400	400
Medical	25	25	25
Breakage deposit	50	—	—
Student activities	40	—	—
Insurance**	10	—	—
Total tuition and fees	$845	$745	$745
Less MARAD subsidy (eligible students)	200	200	200
Net Cost	$645	$545	$545

* Tuition fee for out-of-state students is $175 additional per trimester.
** Insurance is a group policy concerning loss of life or limb.

Clothing, Books, and Supplies

Entering students must deposit in trust $800 for clothing, books, and supplies on or before the first day of the first trimester, to be drawn from as needed. Returning students must deposit sufficient funds to maintain a minimum balance of $50 at the beginning of each trimester. Any unexpended balance in the account will be returned to the student at the termination of enrollment.

Total Costs

	Annual Fees	*Total Fees (excluding·clothing and books)*
In-state		
subsidized student	$1,735	$6,940
nonsubsidized student	2,335	9,340
Out-of-state		
subsidized student	2,260	9,040
nonsubsidized student	2,860	11,440

FINANCIAL AIDS

Financing should not be a barrier to attendance at Cal Maritime. Loans, grants, scholarships and part-time employment are available to those who demonstrate need for assistance and are U.S. nationals. Often, "packages" of two or more kinds of aid are offered to eligible applicants.

Eligibility, unless otherwise noted, is based upon need as determined annually by analysis of the Parents' Confidential Statement (PCS). Applicants should submit the PCS, available from high school counselors and the Academy, and the CMA Financial Aid Application, available from the Academy, prior to March 1, for consideration. Later applications will be accepted if funds are available.

Loans

The National Direct Student Loan is a federally funded long-term loan, repayable beginning nine months after graduation at three percent annual interest. The loan is interest-free until the end of the grace period.

The Federally Insured Student Loan, made by private lenders, and guaranteed by the federal government, is repayable beginning nine months after graduation at 7 percent annual interest. The interest may be paid by the federal government until the end of the grace period, if adjusted family income is under $15,000 per year. This loan requires a separate application, available from the Academy. Application can be

217

made at any time during the year; processing takes approximately two months.

The California Maritime Academy Midshipmen's Loan Fund provides short-term tuition loans at five percent annual interest. Loans must generally be repaid prior to the end of the trimester in which funds are advanced. It, too, requires a separate application, available at the Academy.

Student loan programs are also subsidized by a number of service organizations, such as the California Maritime Academy Foundation, the Propeller Club of the United States, the Society of Port Engineers, and individuals. Contact the Academy for further information.

Grants

The Basic Educational Opportunity Grant is a federal grant for students with exceptional need. A separate application, available from high school counselors and the Academy, must be submitted directly to the federal government.

The Supplemental Educational Opportunity Grant is a federally-funded grant for students with exceptional need. It is offered when other aid is not adequate to meet a student's costs.

The College Opportunity Grant, for entering freshmen who are California residents and who demonstrate exceptional financial need is awarded by the California State Scholarship and Loan Commission. Applications and instructions are available from high school counselors. Application deadline is mid-December for the following academic year.

Employment

The College Work-Study Program, funded by the federal government, provides part-time jobs on the campus. Midshipmen generally work from five to fifteen hours per week at an average wage of $3.00 per hour.

Scholarships

California State Scholarships in the amount of $500 are awarded annually be the California State Scholarship and Loan Commission to California residents, based upon need and academic achievement. Applications are available from high school counselors and the Academy. Deadline is mid-November for the following academic year.

Privately administered scholarships, generally awarded on the basis of need and academic achievement, are offered by many service organizations. Information is available from high school counselors and public libraries. Application is made directly to the donor.

Veteran's Educational Assistance

Cal Maritime is approved for student assistance by the Veteran's Administration. Students should apply to the local office of the VA in their region for assistance and information.

CURRICULUM

The Major

Students at the Academy major in either nautical industrial technology or marine engineering technology. Bachelor of Science degrees are awarded in these two fields. These major programs are organized to reflect the division of labor and responsibility found on vessels of the Merchant Marine, Navy, Coast Guard, and other marine services and industries. The traditions, customs, and regulations of most maritime nations dictate that ship's crews will be divided into the major departments of deck and engineering, and other departments composed of cooks and stewards, pursers, radio officers, doctors, etc.

Nautical Industrial Technology Major (NIT)

The student who aspires to a career as a licensed deck officer majors in nautical industrial technology. This title is used for the deck program because the deck curriculum embodies the two major aspects of industrial technology programs taught at other colleges; namely, a technology concentration and a management concentration. For the NIT program the technology concentration consists of seamanship, navigation, ship operation, cargo handling, and nautical rules of the road. These the deck officer must master to meet his immediate responsibilities as a mate. A mate is also a manager aboard ship. Mates rise to the position of captain or master, who is the commanding or managing officer of the ship. After considerable experience at sea, mates are often given the opportunity to serve ashore in a shipping company or related maritime industry in a management capacity. It is for these reasons that management is the second emphasis in the nautical industrial technology curriculum.

Composition, number, and organization of departments and groups onboard ship vary from nation to nation, industry to industry, company to company, and ship to ship; however, on all ships the master, captain, or commanding officer is the ranking officer aboard to whom the heads and chiefs of the various departments and groups are responsible. Directly under the master in the chain of command are the chief officer, the junior deck officers, and the deck department charged with the navigation, cargo stowage and management of the ship.

Deck licenses issued by the Coast Guard in increasing rank are: third mate, second mate, chief mate, and master. Licenses are further restricted as to waters and vessel tonnage. Nautical industrial

Taking a practice fix.

Ready for inspection.

technology majors will satisfy all the requirements to take the Coast Guard examination for third mate, oceans, unlimited. Further raise in grade is dependent upon the graduate's ability to accumulate sea time and to pass examinations of increasing complexity and difficulty. The higher licenses are issued by the U.S. Coast Guard after satisfactory completion of a written examination and actual seagoing experience, usually one year, in the next grade of license lower to that being issued.

The nautical industrial technology program is designed to give the student the necessary background in navigation, seamanship, cargo handling, and rules for all grades of license up to and including Master. After the requisite experience, it is a comparatively simple matter for an Academy graduate to review his studies, integrate his experience, and successfully undertake the examinations for the successively higher licenses.

Marine Engineering Technology Major (MET)

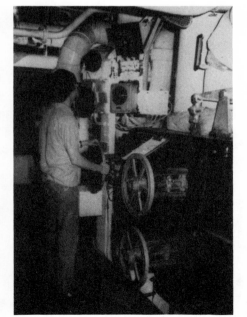

The engineering department of a ship is organized similarly to the deck department with the chief engineer as the ranking officer. The chief engineer on most vessels ranks with the master in salary but, under laws and by tradition, is responsible to the master. Under the chief engineer are the first assistant engineer, the junior engineers, and the engineering department in charge and repair of the vessel's engineering systems.

Engineering licenses issued by the Coast Guard in increasing rank are: third assistant engineer, second assistant engineer, first assistant engineer, and chief engineer. Engineering licenses are further limited as to type of main propulsion machinery, i.e., steam, diesel, etc., and horsepower. Marine engineering technology majors will satisfy all the requirements to take the Coast Guard examination for third assistant engineer, steam vessels, unlimited, with an endorsement as third assistant engineer, motor vessels, unlimited. Further raise in grade is dependent upon the graduate's ability to accumulate sea time and to pass examinations of increasing complexity and difficulty. Higher licenses to engineering officers are issued by the U.S. Coast Guard after satisfactory completion of a written examination and actual seagoing experience, usually one year, in the next grade of license lower to that being issued.

The MET program is designed to give the student the necessary background in marine propulsion systems and the other engineering systems aboard ship for all grades of engineering license up to and including Chief Engineer. After the requisite experience, it is a comparatively simple matter for an Academy graduate to review his studies, integrate his experience, and successfully undertake the examinations for the successively higher licenses.

NIT and MET Majors Compared

At the time of graduation the NET graduate has a specific training that fits him for service as a mate on a ship. As his management experience accumulates and is enhanced by further education, his employment horizons can broaden. On the other hand, the MET student at graduation is fitted not only for service as an engineer aboard a ship, but for employment as an engineering technologist in a wide range of industries ashore. His education is much like that of a graduate of the mechanical engineering technology curriculum offered in other colleges, the difference being an emphasis on power rather than on manufacturing processes. The MET graduate has wider employment horizons than does an NIT graduate, with more jobs to choose from at the time of graduation, and also after a number of years of sailing on his license.

Technology and Traditional Education Compared

Both of Cal Maritime's major curriculums follow the philosophy of technology education. Technology graduates are prepared to be doers in current operations. On the other hand, traditional educational programs in engineering or management give students a background in sophisticated theory for the design and devisement of the engineering devices or systems or the management operations of the future. In traditional programs, knowledge of current techniques and operations must be learned through on-the-job training. The technologist studies theory with lesser mathematical sophistication and with more attention to concepts. He spends a great deal of time studying and operating current equipment. The technology graduate has skills the traditional graduate does not have, skills his employer can put to immediate use. On the other hand, he lacks the sophisticated theoretical knowledge of the traditional graduate. Both kinds of graduates are needed in the modern industrial system. They work together to

Practice lifeboat drill.

make it go.

That the technology graduate has not been exposed to sophisticated theory and methods, does not preclude his or her getting this exposure at a later date through further education. Capable technology graduates are welcome in traditional engineering and business administration graduate programs. It is true that they will have to take some prerequisite theoretical undergraduate courses to get started, but the transition is not very difficult for the capable student. When he or she has acquired the masters degree, they know firsthand how it is done and have full background for the design and innovation of future systems. This is good equipment for the person who aspires to top-level positions.

The Length of Maritime Academy Programs

One who is familiar with other college programs notices that maritime academy programs are longer than those at other colleges. California Maritime Academy has this characteristic in common with other maritime academies. The elapsed time to earn a bachelor's degree is the same, four calendar years, at the academies and at traditional colleges. However, students are in attendance at an Academy for eleven months a year, while at traditional colleges students are in attendance for nine months each year. The Academy midshipman is, therefore, in school for more months and earns more units of academic credit for his Bachelor of Science degree than does a student at other colleges.

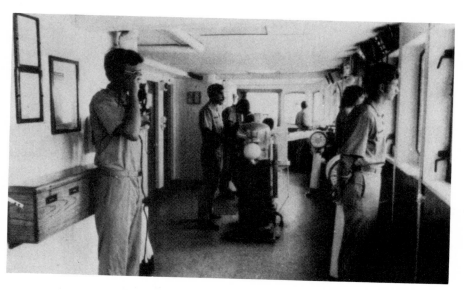

Practice on the bridge.

The reason for this longer program at academies lies in the fact that virtually no on-the-job training is available to the new third mate or third assistant engineer aboard a merchant vessel. The new officer must be able to assume full responsibilities of operation, watch, and equipment maintenance from his first hour on a ship. Preparation for these responsibilities comes from the sea training cruises where the

222

midshipmen operate the school ship under the monitoring and instruction of licensed officers, who are also instructors for the midshipmen in the campus classrooms and laboratories. The total instructional program is thus efficiently integrated to cover all of the theory, equipment familiarization, and operating skills that the new officer must have.

Electives and Options

In order to round out the academic program, maritime-related elective courses are offered by the Academy. Each curriculum requires eight semester units of these electives. Superior students may take an additional eight units as overload, for a total of sixteen, to build a concentration in some speciality. Such a concentration will be called an option, and its completion, together with its title, will be noted on the student's academic record at the time of graduation. Transfer students, because of courses completed elsewhere, at times find themselves with free space in their class schedule. It would be well for them to use this free time to take elective courses toward the eight extra credits required to complete an option. Some courses taken at other institutions, if judged the equivalent of the elective courses in the lists below, may be used to meet elective and option requirements of Academy programs.

Seven options have been developed. They are:

1. Marine transportation;
2. Marine business management;
3. Maritime specialties;
4. Instrumentation and automation;
5. Ocean technology;
6. Naval architectural technology;
7. Nuclear technology; and
8. Elective selection without a specific option

Students may prefer to select elective and overload courses to strengthen and supplement required programs without meeting the requirements of a specific option. In selecting elective courses, prerequisites should be met. If a student wishes, he may take all eight semester units of required elective in one option.

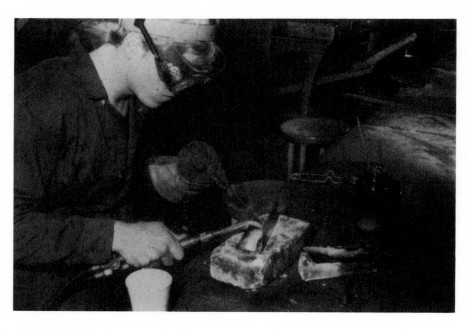

Sea Training

The sea training portion of the curriculum is divided into three training periods of approximately twelve weeks each. During the training periods the midshipmen put the skills and knowledge they have been taught in the classroom to the ultimate test: actual practice. The entire operation of the 491-foot, 7987 gross-ton TS *Golden Bear,* is under management entirely by students, with faculty licensed officers acting in an advisory capacity only. The faculty officers also grade the midshipmen for the degree of professionalism with which they accomplish an assigned task or duty. The fourth class does the more elementary tasks, while the first class performs all the duties of ships' officers, from loading the ship's cargo of provisions and lighting of the boiler plants, to navigating and providing power for the ship to visit exotic ports anywhere in the world.

STUDENT BODY ORGANIZATION AND ACTIVITIES

Corps of Midshipmen

For purposes of organizational training and the further development of a sense of self-discipline, the student body is organized into a Corps of Midshipmen.

A quasimilitary routine is followed, and the midshipmen wear a functional, standard merchant marine khaki uniform most of the time (they wear dress blues, tropical whites, etc. at other times).

The entire student body is under the direction of the Office of the Commandant of Midshipmen, who is responsible for the conduct, welfare and morale of the corps. The Commandant's office is currently under expansion in order that student services will include a sensitive personal counseling program for those who need it.

The corps is divided into eight divisions, which are commanded by students of the senior class, called first class at the Academy. They in turn are responsible to the Corps Commander and his Executive Officer. Such training is maintained in order that the young men and women of the Academy may experience firsthand the chain of command interdependence found on all merchant ships and may gain firsthand experience in the management of personnel and leadership.

Daily Schedule

A midshipman's daily routine begins at 6:30 A.M., followed by breakfast and morning colors formation at 8:00 A.M. After colors, midshipmen have scheduled classes and laboratories in Academy laboratories or aboard the TS *Golden Bear* until 4:00 P.M.

The late afternoon hours are a midshipman's free time, and are normally devoted to varsity athletics, intramurals, club meetings, library study, or some form of extracurricular recreation. Following the evening meal at 5:30 P.M., the rest of the evening is generally spent in studies or liberty. However, to afford educational opportunities not possible during the regular hours, some elective and out-of-sequence classes are scheduled at 4:30 P.M. and 7:00 P.M.

Leave and Liberty

All midshipmen are granted approximately two weeks leave during the Christmas holiday, three days over Thanksgiving, one week in the spring, following the training cruise, and three weeks in July and August, at the conclusion of the academic year.

At the conclusion of classes at 4:00 P.M., the first class is granted liberty until 2:00 A.M. Second class liberty commences at 4:30 P.M. and ends at 1:00 A.M. Third class liberty begins at 4:30 P.M. and ends at midnight.

Fourth class midshipmen are granted only Wednesday evening liberty in addition to weekend liberty during the first two trimesters at the Academy, in order that healthy study habits can be formed. Midshipmen on the twenty-four hour watch section or on conduct restriction must remain on the campus at all times.

The Commandant's office grants sick leave or emergency leave to any midshipmen when circumstances warrant, and also grants special leave or liberty for extracurricular activities and special events.

Medical and Dental Care

A medical assistant is available for first aid. Additionally, the services of a contract physician are available. During the sea training, a licensed physician is aboard the training ship. Hospitalization is available at the U.S. Public Health Service hospital in San Francisco.

All of the foregoing are furnished at no cost to the student. However, should injury or illness occur while the student is away from the Academy, except as noted, any expense for emergency treatment or transportation to the U.S. Public Health Service hospital must be borne by the student. While enrolled at the Academy a student is covered by a $5,000 loss of life or limb policy. When away from the Academy as a member of an athletic team, a student is covered by a $1,500 accidental death or medical hospital indemnity-accident insurance policy. Dental treatment is furnished at the U.S. Public Health Service hospital in San Francisco.

EXTRACURRICULAR ACTIVITIES

On Campus

A center for student activities and recreation has been completed and includes a coffee mess, pool table, table tennis, television lounge, study, typing cubicles, and reading, lounge, and meeting areas. A television lounge is provided in the dormitory for student use during free time; there are reading lounges in the campus library. There is also a canteen in the dormitories for the purchase of small items.

The gymnasium, exercise room, and pool are kept open in the evening for use by students.

A variety of clubs and special interest groups exist with full endorsement of the Academy and financial support in the form of equipment and material. The monthly student newspaper, the *Binnacle,* and the annual yearbook of the graduating class, the *Hawsepipe,* are two activities receiving Academy financial support, and students interested in journalism are encouraged to participate in their publication. Students are also encouraged to take the initiative in forming new clubs and special interest groups with faculty advisement.

There are no fraternities or sororities sanctioned by the Academy, nor does the Academy take any official stand, pro or con, on the existence of such organizations.

On-campus social and recreational events currently include basketball, water polo and soccer games, postgame dances, and weekend picnics for alumni and other Academy-related groups. Considerable expansion of the on-campus cultural and social calendar takes place in the newly completed 500-seat auditorium.

The honor guard and drill team, sponsored by the Academy through the Department of Naval Science, practice on campus, but participate in parades and other competitions throughout the state, winning many honors and awards yearly.

Off Campus

The San Francisco Bay area is world renowned for the variety and richness of its religious, cultural, educational, and entertainment activities. The student will have no difficulty in finding pursuits to fill his or her leisure time.

Driving time from Vallejo in the extreme north of San Francisco Bay to San Jose in the extreme south is less than two hours. Included within this driving range are social, educational, and cultural events, conducted on campuses of the University of California and California State Universities, as well as several private universities and community colleges, that are free or at reduced rates for Cal Maritime students. Also within this range are the San Francisco and Oakland symphonies; the Oakland Center for the Performing Arts; the American Conservatory Theater, widely recognized as one of the finest theater companies in the world; the San Francisco Opera and Ballet Company; various public and private theaters offering legitimate theater, musical comedy, and pop concerts; countless public and private museums and art galleries; and zoological and botanical gardens. Many of these functions are free or at reduced rates for Cal Maritime students.

Cal Maritime teams participate in intercollegiate soccer, water polo and basketball and soccer, and other intercollegiate sports events can be attended by Cal Maritime students at reduced rates. The San Francisco Bay area is home for major league baseball, football, basketball, and ice hockey, and Cal Maritime students are eligible for student rates for these events wherever offered.

For outdoor lovers, the Academy and the California Maritime Academy Foundation operate several power and sailing boats and yachts used for cruising and fishing around the Bay, Sacramento and San Joaquin Rivers and Deltas, and near California coastal waters of the Pacific. Midshipmen are welcome on these cruises as crew members, operators or guests, which is an excellent opportunity to practice and polish techniques of small-boat handling and seamanship or to just relax.

The California Maritime Academy Foundation also uses its yachting fleet to give an opportunity to disadvantaged youth to learn something about boating, and midshipmen are invited to participate in this program.

Other outdoor activities in the area include picnicking, hiking, horseback riding, surfing, scuba diving, and so on, in many national, state, and regional parks; all within the two-hour driving time. Just to mention a few, there are: Golden Gate Park in San Francisco; Tilden Park in Berkeley; Muir Woods Park in Marin County, where one of the most inspiring stands of coastal redwoods, *sequoia sempervirens,* is to be seen; Mount Diablo Park in Contra Costa County; Angel Island recreation area in San Francisco Bay; Stinson and Bolinas Bay beaches in Marin County, and San Francisco Beach in San Francisco. Admittance to these facilities is gratis or nominal.

Off-campus events are, of course, myriad; however, the Academy, through the student council, sponsors the annual Ring Dance at one of the large hotels or private clubs in the Bay area.

Religious Practice

The Academy has no official stand on religious practice, nor are any formal religious observances conducted on campus or the training ship with mandatory attendance by students. However, since every major religion and many religions not commonly considered major are active within the two-hour driving time from the campus, no student should unwillingly suffer from lack of spiritual guidance and fellowship.

The Protestant, Roman Catholic, Eastern Orthodox, and Jewish faiths all have congregations in Vallejo or environs, ethnic Chinese and Japanese Buddist, Taoist, and Zen societies and assemblies, as well as churches, exist in San Francisco and other parts of the Bay area, and an Islamic Center and Moslem Mosque are located in San Francisco. There are numerous Spanish-speaking Protestant and Roman Catholic congregations scattered throughout the San Francisco Bay area and in the Sacramento and San Joaquin Valleys. There is a variety of Eastern Orthodox congregations in the Bay area with liturgies in the Greek, Russian, Armenian, Serbian, and Arabic languages. Virtually all congregations in the Bay area have their own social programs in which the student is free to participate; however, several religious clubs are in existence on campus whose members attend services as a group and hold prayer meetings and breakfasts during the year.

Industrial Contact

Whenever the student has free time, i.e., weekends, holidays, and vacations, he is urged to go to sea aboard various west coast vessels for a period of firsthand observation of the real-world operation of merchant vessels. These trips are organized between the Academy and the many steamship companies in the San Francisco Bay area. While aboard the vessel, each student serves as an observer-cadet under the direction and assistance of the ship's officers. Voyages for the students may be just a few days up and down the coast, or they take the student to Alaska, Hawaii, or Japan.

ACADEMIC PROGRAM

Accreditation

The California Maritime Academy is accredited by the Accrediting Commission for Senior Colleges and Universities of the Western Association of Schools and Colleges as a candidate of the Commission.

In addition to institutional accreditation, the college is pursuing accreditation of its two degree programs. The Engineers Council for Professional Development granted the marine engineering technology program candidate status in its accreditation process. This status is subject to annual review.

The National Association of Industrial Technology, not having a candidate status, has had a representative pay a consulting visit to the campus for the purpose of observing and reviewing the nautical industrial technology program.

Degrees

The Bachelor of Science in nautical industrial technology or the Bachelor of Science in marine engineering technology is conferred upon midshipmen who have successfully completed the academy's program of instruction and the applicable U.S. Coast Guard license examination.

Licenses

Midshipmen who meet the physical and educational requirements of the U.S. Coast Guard examination are licensed as third mates or third assistant engineers, and are qualified in these capacities to serve aboard any American flag ship.

School Year

The academic year is divided into three trimesters. The fall and spring trimesters are approximately seventeen weeks each, and the winter trimester is approximately twelve weeks in length. A brief recess follows each trimester.

ACADEMIC STANDARDS

Grading System

The letter grading system, with corresponding grade points, is used to indicate the caliber of the student's work. The scholastic significance of the grades is:

Letter scale		Grade points
A	outstanding	4
B	excellent	3
C	average	2
D	minimum passing	2
D	minimum passing grade	1

A midshipman is expected to maintain a grade point average of 2.0 (C) or better to continue in good standing. To be eligible for the B.S. a student is expected to complete the program with a grade point average of 2.0 (C) or better.

CAREER OPPORTUNITIES IN TODAY'S MARITIME INDUSTRIES

Career opportunities are many in America's maritime industries. The vast production of U.S. industry cannot be consumed by the domestic market alone, nor can our factories or refineries produce without importing essential raw materials. Seven-tenths of the globe consists of water, and since foreign trade depends largely upon ships, ocean shipping becomes of greater importance to the American economy than ever before in our history. An active merchant marine and the knowledge required to operate merchant ships is essential for the commerce and defense of our nation.

The demands of commerce have radically changed the complexion of the merchant marine from the days of the small, slow, lumbering sailing ships to vessels with a capacity of hundreds of thousands of tons that travel at speeds never previously believed possible.

These larger, faster ships demand crews highly trained in the most modern marine technology known to man. In addition to the traditional skills a mariner must master, he must also be skilled in such fields as electrical engineering, electronic systems, marine nuclear science, marine ecology, meteorology, oceanography, marine transportation management, computer technology, and intermodal transportation concepts.

Within the last few decades the owners and operators of offshore production, research, exploratory, and service vessels have exhibited a keen interest in Cal Maritime graduates of both majors for employment at sea and as marine managers ashore. This trend is expected to continue into the future with development of the deep ocean resources.

Today's maritime industry is a global enterprise possessing limitless opportunities for the ambitious both afloat and ashore.

Today Cal Maritime graduates can be found employed in virtually every capacity of the maritime and related industries, from marine insurance to naval architecture. In view of the wide range of knowledge required of a merchant marine officer in today's maritime industry, career opportunities for academy graduates have increased considerably in many fields of endeavor, and today's graduates are highly employable.

Cal Maritime graduates' beginning salaries for shore jobs are among the highest for any college graduates. Net income for initial seagoing jobs is from one and one-half to more than two times the national average for college graduates with Bachelors or Masters degrees.

For further information about the California Maritime Academy write to: Admissions, California Maritimy Academy, P.O. Box 1392, Vallejo, CA. 94590.

APPLICATION

Request an application for admission by writing or telephoning the Admissions Office, California Maritime Academy, P.O. Box 1392, Vallejo, California 94590, telephone: (707) 642-4404.

Submit the application to the Registrar, who will respond with specific information regarding additional documents required. Specifically, these are:

1. Test scores of entrance examinations;
2. One official copy of high school transcript;
3. Two official transcripts of all college work attempted;
4. Three copies of birth certificate from the issuing agency;
5. Three letters of recommendation, at least one of which is from a high school or college counselor or principal;
6. A statement of residence (will be mailed on receipt of application); and
7. U.S. Coast Guard physical (authorization will be mailed on receipt of application).

Applications are processed, and acceptance letters are issued as soon as the applicant's file is complete. Application prior to April 1 is advised. Late applications will be considered if space is available.

Texas Maritime Academy,
Moody College of Marine Sciences
and Maritime Resources

The Moody College of Marine Sciences and Maritime Resources is located in Galveston, Texas. Its purpose, in conjunction with other colleges and programs of Texas A&M University, is to provide academic instruction and extension services, as well as to conduct research commensurate with the increasing importance of marine affairs to coastal Texas. Moody College also coordinates all of the University's programs in the Galveston area.

Moody College, created in September of 1971, consists of the Department of Marine Sciences, the Texas Maritime Academy, and the Galveston Coastal Zone Laboratory. Currently comprising five marine degree programs, the undergraduate curriculum of the College will be expanded in the future to cover the full spectrum of marine subjects. All present degree programs lead to the Bachelor of Science degree from Texas A&M University.

Classes are held on Mitchell Campus, Pelican Island, as well as at Fort Crockett on Galveston Island. The TS *Texas Clipper,* training ship of the Academy, serves the students as classroom and dormitory both ashore and at sea. The 15,000 ton converted cargo and passenger liner is berthed at Pelican Island during the regular school year, and is manned by cadets each summer on a training cruise. The ship may be visited on Saturday and Sunday afternoons on Pelican Island during the regular school year.

The location of the College in Galveston affords students an opportunity to utilize facilities of the maritime industry ashore and afloat, as well as to benefit from field research and instruction in the bay and in estuarine and nearshore waters.

HISTORY

Texas Maritime Academy, the nation's fifth state maritime academy, operates academically as a division of Moody College of Marine Sciences and Maritime Resources of Texas A&M University.

The Academy was created in 1962 under an agreement between the State of Texas and the United States Maritime Administration. Texas A&M University, acting for the state, receives federal support for the Academy in the form of a training ship, annual appropriations for ship maintenance, $75,000 per year in operating funds for the Academy's programs, and a subsidy provision of $50 per month for a total of 140 eligible cadets.

ACADEMY PROGRAMS

The programs of Texas Maritime Academy, marine engineering and marine transportation, have the option to prepare students to become merchant marine officers, with U.S. Coast Guard licensing. License-option students must complete three training cruises for a total of six months at sea to become eligible for the Coast Guard licensing. They may qualify as third mates (marine transportation, marine biology or marine sciences) or third assistant engineers. License option students are required to join the Corps of Cadets of Texas Maritime Academy. Nonlicense students may join the Corps, but are not required to do so.

The U.S. Merchant Marine is composed of all privately-owned American flag vessels, including oceangoing ships and the smaller craft concerned with work in local rivers, harbors, and coastal areas, as well as the personnel required to operate the vessels and support activities. Merchant marine officers are responsible for the safe operation of the merchant vessels. The Coast Guard issues licenses to officers after they have fulfilled the requirements of training and experience, and have a written examination.

Following graduation and licensing, the new third mate or third assistant engineer may join a ship as a fully qualified junior officer. He or she will be responsible for the safe navigation of the vessel, loading and discharging of cargo, vessel maintenance, and shipboard safety, as a third mate, or, as a third assistant engineer, be responsible for the maintenance and operation of all machinery aboard ship. This includes propulsion machinery, auxiliary machinery, electrical systems, refrigeration machinery, and air-conditioning systems.

After one year of shipboard experience, the examination for a second mate's or second engineer's license may be taken; after serving a year as a Second, the young officer is qualified to sit for the chief mate's or first assistant engineer's license. The final step, after additional service and experience, is licensing as master or as chief engineer.

Texas Maritime Academy offers degree programs in marine engineering and marine transportation with the license option. There are also license-option programs with the major in marine sciences and in marine biology.

Summer School at Sea

Recent high school graduates who enroll in Texas A&M University as freshmen may earn their first six college semester hours aboard the TS *Texas Clipper* during the annual summer training cruise. As new Academy cadets, they choose two courses for six semester hours credit from offerings in English, history, and mathematics. In addition to daily classes, they also are responsible for assisting the ship's crew in maintaining and operating the vessel during the training cruise.

For those students interested in marine engineering, firsthand experience with the ship's operation and power plants is available. For those interested in marine transportation, there is opportunity to work on the bridge or on deck under the supervision of a licensed merchant marine officer. The program also allows the potential merchant marine officer to determine if his or her initial attraction to the sea is one that can be directed toward a career in the maritime service through the college curriculum.

The *T.S. Texas Clipper*

ACADEMICS

Department of Marine Sciences

The Department of Marine Sciences directs the degree programs of marine biology, marine sciences, and maritime systems engineering. The license option also is available in the marine sciences program, whereby a student may become eligible for licensing by the U.S. Coast Guard as a deck officer in the merchant marine. Maritime systems engineering options include hydromechanics, ocean engineering, and coastal structures.

The Department of Marine Sciences, now housed at Texas A&M University's Fort Crockett facility on Galveston Island, is devoted to year-round research and instruction, both graduate and undergraduate, in various disciplines related to the marine sciences.

Galveston Coastal Zone Laboratory

The Galveston Coastal Zone Laboratory is the third component of Moody College of Marine Sciences and Maritime Resources, and serves as its research arm. Short-term research pertinent to the Galveston coastal area and applied research in ocean, bay, and estuarine systems are conducted at the Coastal Zone Laboratory. Research activities have included oyster mariculture, use of offshore oil rigs for oceanographic engineering, distribution of Blue crab in experimental temperature gradients, development of a pilot oyster hatchery, studies of shrimp, and other biological surveys. Agencies initiating research have included Houston Lighting and Power Company; National Science Foundation; National Maritime Research Center; Health, Education, and Welfare; Bureau of Land Management; and National Sea Grant Program.

DEGREE PROGRAMS

Marine Biology (MARB)

The marine biology program leads to the Bachelor of Science degree in Marine Biology. The program is structured for training the student in biological disciplines concerned with coastal and marine environments. The program further provides a focus for marine biological education in the coastal zone with required field excursions. Preparation in the sciences is recommended.

Marine Engineering (MARE)

The marine engineering program leads to the Bachelor of Science degree in Marine Engineering; it has a license-option program, whereby a student can qualify to sit for the U.S. Coast Guard license examination for third assistant engineer, steam and motor vessels, oceans, unlimited. Engineering theory and practice are coordinated by relating classroom study to the student's practical experience aboard the *Texas Clipper,* as well as by visits to ships and maritime industries in the Galveston-Houston port area.

Marine engineering, closely related to mechanical engineering, em-

phasizes the design, operation, and maintenance of marine power plants and associated equipment. Thorough preparation in mathematics, sciences, and basic and applied engineering is recommended for students pursuing this degree program.

Marine Sciences (MARS)

The marine sciences program, with a general approach in science and humanities the first two years and a specialization in marine disciplines the last two years, leads to the Bachelor of Science degree in Marine Sciences. The program is designed to train students for employment in marine areas concerned with fisheries, biology, oceanography, and ecology, as well as employment as secondary school teachers of marine science subjects. Graduates may also engage in further study in biology, marine biology, marine fisheries, oceanography, marine and coastal ecology, and marine resources and coastal zone management.

In cooperation with the Department of Marine Transportation, Texas Maritime Academy, an option for U.S. Coast Guard license as third mate is offered. The license-option student pursues the marine sciences degree program, but is a member of the Corps of Cadets. He or she must meet Academy qualifications. The student may additionally be eligible for a U.S. Naval Reserve Commission as ensign upon graduation. Graduates would qualify for careers as licensed third mates aboard research vessels, exploration vessels, and merchant ships, or they could elect to continue their studies in pursuit of a graduate degree.

Marine Transportation (MART)

The marine transportation curriculum provides a basic licensing and degree program for deck officer candidates, including option pro-

grams in marketing and management. The graduate, with a Bachelor of Science degree in Marine Transportation, will have completed a program that combines the humanities and sciences with marine subjects, in order to meet the present and future needs of the maritime industry at sea and on shore.

Theory and practice are integrated by relating the scholastic efforts with the sea training periods on the *Texas Clipper,* and with visits to ships and maritime industries in the Galveston-Houston port area. The student who successfully completes the license program will be qualified to sit for the U.S. Coast Guard license examination for a federal license as a 'third mate, steam and motor vessels, oceans, unlimited.

Maritime Systems Engineering (MASE)

The maritime systems engineering curriculum concentrates on fundamental engineering design in combination with humanities, sciences, and various marine subjects. A general core of courses in humanities, sciences, and engineering during the freshman and sophomore years provides a foundation for specialization in the options during the junior and senior years.

The program is aimed at training students for employment in any marine-oriented engineering field. Students are accepted as entering freshmen or as transfers from engineering, mathematics, or physical science programs at junior and community colleges. Some transfers are accepted from four-year institutions when the students desire to concentrate their education in the coastal zone. With the proximity of the Ports of Galveston and Houston, there are field trips and guest lectures from the local marine industries.

ADMISSION

Undergraduate Admission

Texas A&M University is a coeducational institution that admits all qualified applicants on an equal basis without regard to sex, race, creed, color, or national origin.

Applications for admission are welcome at any time. Those who meet the standards will be admitted as long as space is available, until the last day for enrollment during the session requested.

Applications for admission to the Moody College of Marine Sciences and Maritime Resources should be addressed to the Office of Admissions, Texas A&M University, College Station, Texas 77843. Completed application forms, to be submitted to the Office of Admissions, must be accompanied by transcripts of credit, if the applicant is entering directly from high school. A student attempting to transfer from another college or university must have two complete official transcripts from each college or university attended.

An applicant must have graduated from a properly accredited secondary school with a minimum of sixteen units (credits) that are acceptable to the University for entrance purposes. Students who have a superior high school record and who wish to enter higher education without graduating from high school may apply provided they present the desired sixteen credits as outlined below, score at least 1100 on the SAT, and rank in the highest quarter of their high school class. In addition, they must be recommended by their high school principal. A personal interview with the Director of Admissions will be required prior to admission.

The unit requirements for admission to the University are designed to insure adequate preparation for the various curriculums offered by the University. To give deserved recognition to proven ability as reflected by high academic achievement, a student may enter the University even though he has not completed all the required high school subjects as listed below. An applicant who ranks in the highest quarter of his class and scores a minimum total score on the CEEB Scholastic Aptitude Test of 1000 may be granted admission with credit deficiencies.

Entrance Credits

The sixteen acceptable entrance credits that a student should have for admission (with exceptions indicated where applicable) are as follows:

Subject	Units of credit required	Remarks
English	4	Required of all students. Two units in a single foreign language may be substituted for one unit in English.

Social science	2½	Required of all students.
Mathematics		
algebra	2	Required of all students.
plane geometry	1	Required of all students.
Science	2	Required of all students. It is preferred that these two units include biology, chemistry, or physics.
Electives	4½	Recommended from the following subject areas: foreign languages, mathematics, science, social science, speech. Not more than three vocational units may be submitted as electives. Applicants are strongly advised to include at least ½ unit elective in advanced mathematics.
Total	16	

Tests Required of New Students

Texas A&M University requires certain College Entrance Examination Board (CEEB) tests as part of its admission requirements for students entering college for the first time. Results of these tests are used for admission, counseling, and placement purposes.

The following tests are required: Scholastic Aptitude Test (SAT), English Composition Achievement Test, Mathematics Achievement Test (Level I or II). CEEB offers these examinations at testing centers throughout the United States and in major cities of many foreign countries. Testing dates, locations, and fees are described in a bulletin that may be obtained by writing the College Entrance Examination Board, Box 1025, Berkeley, California 94701. The CEEB reporting number for the Moody College of Marine Sciences and Maritime Resources is R6835.

It is the applicant's responsibility to arrange to take the examinations. Arrangements are to be made directly with the College Entrance Examination Board, not through Texas A&M University.

The minimum test scores for applicants who have never attended another college or university are stated in terms of a total score on the College Entrance Examination Board's Scholastic Aptitude Test (SAT). As reported by the board, the total of the verbal and mathematical scores becomes a criterion for admission.

Entering freshmen must meet the following SAT minimums:

Standing in high school graduating class	Minimum total score acceptable for admission
Highest quarter	800
Second quarter	800
Third quarter	900
Fourth quarter	1000

Early-Decision Program

In order to recognize and reward superior academic performance, Texas A&M has instituted an early-decision program that permits a student to apply for admission after he has completed his junior year of high school. To be eligible for this program a student must rank in the highest quarter of his class and score at least 1000 on the SAT. Students who desire to apply under these provisions may submit their application for admission at the completion of their junior year of high school. A list of courses they will be taking during their senior year must be included with the transcript. Acceptance will be conditional until the student has satisfactorily completed the courses in progress for the senior year and graduated from high school.

Application for Admission to Texas Maritime Academy

In addition to general admission requirements, applicants who wish to participate in the federally subsidized license-option program of the Texas Maritime Academy must be citizens of the United States and physically fit. The physical requirements for a deck cadet include uncorrected vision of at least 20/100 in both eyes, correctable to at least 20/20 in one eye and 20/40 in the other; for an engineering cadet, the corrected vision must be at least 20/30 in one eye and 20/50 in the other. The color sense will be tested by means of a pseudoisochromatic plate test, but any applicant who fails this test will be eligible if he can pass the "Williams" lantern test or equivalent. Thirty-five federal subsidies exist for eligible entering Texas Maritime Academy cadets each year.

License-option students must also complete three training cruises, for a total of six months at sea, to be eligible to take Coast Guard examinations for third mate (marine sciences, marine biology and marine transportation) or third assistant engineer (marine engineering). License-option candidates must join the Corps of Cadets of Texas Maritime Academy, and annually, thirty-five cadets are eligible for $50 per month federal subsidy for uniforms and textbooks. Nonlicense-option students may choose to join the Corps, but are not required to do so.

Admission of Transfer Students

Admission may be granted to undergraduate students who begin at other colleges and who meet admission requirements. Applicants may not disregard their academic record at any other institution, but must be eligible to return to the school from which they seek to transfer. Formal application must be made and submitted with two official transcripts from each school previously attended.

The applicants must achieve an overall grade point ratio of at least 2.00 (C average on a four point scale), and they must have surpassed that average for the most recent two semesters of attendance, if they have completed that much college work. (A twelve-week summer session with a normal load of coursework is considered a full semester.)

Transfer applicants who have attempted eighteen semester hours or less must achieve the 2.0 standard and also comply with requirements for entering freshmen. High school records, college records, and test results will be used to determine admission status. Either the CEEB Scholastic Aptitude Test (SAT) or the American College Testing (ACT) program will be acceptable for determination of transfer admission status.

On the basis of credentials submitted, credit will be given for satisfactory work completed at another accredited institution, so far as the work completed is equivalent in character and extent to similar offerings of Texas A&M Univesity. Transfer credits are Provisional and may be cancelled at any time if the student's work at Texas A&M is unsatisfactory.

Courses in a subject that are more elementary than the beginning required courses in that saame subject area of a student's chosen curriculum at this University will not apply toward satisfying the degree requirements of that curriculum.

Special Admissions Admission by Individual Approval

An undergraduate applicant who has not recently attended school, and who cannot satisfy the entrance requirements in full, may be admitted if the applicant applies on the official entrance forms and can furnish evidence that his or her preparation is substantially equivalent to that required of other applicants, and that he or she has the ability and seriousness of purpose necessary to pursue studies with profit and to the satisfaction of the University.

EXPENSES

The Board of Regents of the Texas A&M University System, in recognizing the regional character of Texas Maritime Academy, has ruled that both in-state and out-of-state students pursuing the license option program and who are enrolled in the United States Maritime Service will pay tuition at the rate of $5.00 per semester hour, with a $60.00 per semester minimum. Students who are residents of the State of Texas and who are pursuing one of the non-license option programs will pay tuition at the rate of $4.00 per semester hour

($50.00 per semester minimum). Out-of-state students pursuing a non-license option will pay tuition at the rate of $40.00 per semester hour.

Below is an estimate of expenses for students who are pursuing the license option program and are registered for 18 semester credit hours. It should also be noted that students who are participating in the Corps of Cadets program will have an initial outlay for uniforms of approximately $200. All fees listed are strictly approximations and are subject to change because of economic conditions and/or legislative requirements.

	Fall Semester	Spring Semester	Summer Cruise
Tuition ($5.00 semester hour)	$ 90.00	$ 90.00	$ 60.00
Student Services	18.00	18.00	18.00
Board	426.00	426.00	426.00
I. D. Card	3.00		
Room	375.00	375.00	185.00
Cruise			310.00
Laundry			30.00
Total	$912.00	$909.00	$1,029.00

Estimated expenses for resident students not following a license option program are approximately the same while non-resident students should add $630/semester for out-of-state tuition charges. Of course, students not enrolled in a license option program will not be required to pay summer training cruise expenses.

Room, Rent, and Board

All Texas Maritime Academy license-option students are required to pay room and board. Room rent includes heat, light, and cleaning of the corridors, but not the rooms. Rooms are furnished with beds, wardrobes, desks, chairs, and dressers. Students are expected to furnish pillows, blankets, and linens.

Textbooks and Supplies

The cost of textbooks and supplies is approximately $200.00 for the combined fall and spring semesters, but will vary depending upon the course of study and the quality of supplies purchased. The College operates a book store in the Student Activities Building.

Fee Exemptions

Statutory Provisions. Students may be exempt from paying tuition if they qualify as:

1. Highest ranking high school graduate;

2. Veterans who were citizens of Texas at the time they entered service and have resided in Texas for at least the twelve months before the date of registration; eligible dependents of Texas veterans who have resided in the state for at least twelve months immediately preceding the date of registration;
3. Dependent children of firemen and peace officers disabled or killed while on duty;
4. Blind and deaf;
5. Students of other nations of the western hemisphere;
6. Firemen enrolled in fire science courses; or
7. Children of prisoners of war or persons missing in action.

Students may be exempted from paying laboratory fees if they are:

1. Veterans who were citizens of Texas at the time they entered service and have resided in Texas for at least the twelve months before the date of registration; eligible dependents of Texas veterans who have resided in the state for at least twelve months immediately preceding the date of registration.
2. Dependent children of firemen and peace officers disabled or killed while on duty;
3. Blind and deaf;
4. Firemen enrolled in fire sciences courses; or
5. Children of prisoners of war or persons missing in action.

In addition, student services fees are not charged to children of prisoners of war or persons missing in action, or dependent children of firemen and peace officers who have been disabled or killed while on duty. Blind and deaf students are exempt from paying the student services fee and the general property deposit. No student is exempt from paying charges for room and board unless he does not live in campus housing and does not use board facilities.

Board of Regents' Provisions. The Board of Regents has exempted full-time employees of the Texas A&M University System and students registered in absentia from paying student fees.

Claims for exemption from any charges or fees must be supported by evidence sufficient to enable the Registrar to verify the student's exempt status and to determine the duration of the exemption and the fees and charges to which it is applicable.

STUDENT ACTIVITIES

Clubs

Chartered clubs on campus include: Marine Sciences, Outdoor Sportsmen, Surfing, Yacht Club, The Propeller Club and The Diving Association.

Student Publications

Students publish a weekly newsletter: *Channel Chatter,* a yearbook,

The Voyager and a literary magazine, *Seaspray.*

Student Government

The evolving student government of Moody College is embodied in the Student Advisory Committee to the Provost. Members are elected each year on the basis of class and division. The College's seat in the Texas A&M University Student Senate is filled by a student elected from the college at large.

The Student Advisory Committee serves as a direct communications link with the administration on student affairs. It also conducts many programs of service to the students, such as assistance in voter registration.

Intramural Athletics

The intramural program attempts to provide each student with the opportunity to participate regularly in organized activities, according to time and interest. Teams are organized in flag football, basketball, softball, table tennis, volleyball, soccer, and baseball.

Texas A&M University is a member of the Southwest Conference, which includes nine leading universities—the University of Texas, Texas A&M University, Baylor University, Rice University, Texas Christian University, Southern Methodist University, the University of Arkansas, Texas Tech University, and the University of Houston. The intercollegiate program includes football, basketball, track, cross-country, baseball, swimming, tennis, and golf. The Texas Aggies wear the university's colors, maroon and white. Student tickets to all University athletic events are available through the Advisor for Student Activities.

STUDENT SERVICES

Scholarships and Financial Aid

The TAMU Scholarship Program is administered by the Faculty Scholarships Committee. The overall program is designed to encourage and reward scholastic effort on the part of all students, to enable outstanding students to do their best work by removing financial handicaps, and to enable those who might be denied an education for financial reasons to secure an education at Texas A&M University.

In general, there are three types of grants-in-aid available: Valedictory Scholarships and Opportunity Awards, limited to entering freshmen; scholarships designed for more advanced undergraduate students; and fellowships for graduate students. Students at Moody College are eligible to participate fully in all of the scholarship and financial-assistance programs available to colleges of Texas A&M University.

Valedictory Scholarship. This scholarship is offered to a valedictorian who graduates from a secondary school accredited by the Texas Education Agency and qualifies for admission to the University. The successful applicant must earn the recognition by having, among all students, the highest grades, and must be certified to the University throught the Texas Education Agency.

A Valedictory Scholarship will exempt a recipient from payment of tuition during both semesters of the first long session immediately following graduation. When the circumstances of an individual case (usually military service) merit such action this exemption may be granted by the University President for any one of the first four long sessions following graduation from high school.

The Opportunity-Award Program for Entering Freshmen. This annual program provides approximately 400 four-year awards to high school graduates who are capable of outstanding scholastic achievement and who may need financial assistance to attend Texas A&M University. Financial benefits range in value from $400 to $3800; recipients receiving from $100 to $750 each year for four years. Most awards are unrestricted as to course of study or degree objective. Educational Opportunity Grants made available under the Higher Education Act of 1972 are also administered through this program. Graduates of accredited high schools who have not attended another college or university, and who are single, are eligible to apply for an Opportunity Award Scholarship.

Winners are selected by the College Scholarships and Awards Committee on the basis of the applicant's academic record in high school, College Entrance Examination Board test scores, and evidence of initiative, leadership, and other traits of good character. In order for the award to be continued from semester to semester, the recipient must maintain a standard of scholastic achievement and personal conduct satisfactory to the Faculty Scholarships Committee.

Application blanks are available upon request. Requests for additional information and application forms should be addressed to the Director, Student Financial Aid, Room 303, YMCA Building, Texas A&M University, College Station, Texas 77843.

Scholarships for Advanced Undergraduate Students. Scholarships ranging in value from $100 to $1000 are available to students already enrolled in the University. Some of these awards are limited to certain fields of study and to individuals who have attained a necessary academic classification, while others are unrestricted. Each year, recipients are chosen by the Faculty Scholarships Committee in May. The basis of selection is determined by the nature and intent of the award.

Some of these scholarships are given as "rewards for a job well done" and are intended to recognize outstanding scholastic achievement or other meritorious accomplishments.

In addition to the reward type of scholarship, others are made available to outstanding students who must have financial assistance in order to remain in college.

There are also a limited number of college-level scholarships awarded through the Scholarships and Awards Committee of Moody College. These awards are made possible through annual donations from organizations, such as the Women's Propeller Clubs of Galveston and Sabine, and the Women's Organization of the Propeller Club of New Orleans.

Information regarding scholarships for advanced undergraduate students may be obtained from the Advisor for Student Activities, Moody College of Marine Sciences and Maritime Resources, Galveston, Texas 77550.

Employment for Students

Part-time employment of students is coordinated by the Advisor for Student Activities. To become eligible for employment, a student must have been admitted to the University by the Dean of Admissions and have an accepted application on file with the Advisor for Student Activities.

Texas A&M University participates in the College Work-Study Program authorized by the Economic Opportunity Act of 1964.

Loan Funds

The University is participating in both the Hinson-Hazlewood College Student Loan program and the federally insured student loan program. Repayment on the loans begins after graduation. Applications for these loans must be submitted sixty days prior to the time of need. Inquiries should be addressed to the Fiscal Officer, Moody College of Marine Sciences and Maritime Resources, Galveston, Texas 77550.

For students of the Texas Maritime Academy in the license program, the Superintendent's Loan Fund also grants loans of up to $500 to be repaid after graduation.

Other emergency loans are available to all enrolled students. A small service charge is made for these loans. Eligibility is based upon the student's satisfactory record, and the amount of each loan depends upon the student's actual needs.

Vocational Rehabilitation Aid

The Texas Education Agency, through the Vocational Rehabilitation Program, offers assistance for tuition and required fees to certain students in Texas colleges and universities. Eligibility for such assistance is based on permanent physical disabilities.

Application should be made to the Texas Rehabilitation Commission, Room, 309, YMCA Building, Texas A&M University, College Station, Texas 77843, or to the Texas Rehabilitation Commission, 1600 West 38th Street, Austin, Texas 77831.

Counseling

Limited specialized counseling of students is available upon request. Referral by university representatives, parents, or other persons is also possible.

Personal and other problems will be handled in confidence by professional counselors. The Advisor for Student Activities may also call upon other resources of the University in helping students adjust to particular problems.

Housing

New campus facilities were completed in the fall of 1976 and include dormitories, a physical plant, and a student activities building. License-option students are required to live on campus, taking both room and board. Regular students may live off campus, but may pay board. The new Student Activities Center will have food service facilities, for arrangement per meal or per semester.

Besides available housing on the *Texas Clipper* and in the dormitories, students may also choose to live in the community. The Advisor for Student Activities assists students in securing housing or finding roommates.

Health Services

No health facilities or care are available except that TMA license-option students receive care through the U.S. Public Health Service. However, student health insurance is available. For further details, contact the Advisor for Student Activities.

Placement of Graduates

Moody College assists in the placement of its graduates, and fulfill-

ment of this program is a continuing interest of the faculty in each discipline. Active contact is maintained with prospective employers so that graduates will be directed to the best career opportunities.

Application

For further information concerning the Texas Maritime Academy write to: Admissions, Texas Maritime Academy, Moody College of Marine Sciences and Maritime Resources, Galveston, Texas 77550.

The Maine Maritime Academy

The Maine Maritime Academy, chartered by the State of Maine, is accredited by the New England Association of Schools and Colleges. It is an academic member of The Research Institute of the Gulf of Maine (TRICOM), and a member of the American Association of State Colleges and Universities. The Academy is authorized under Federal law to enroll nonimmigrant alien students.

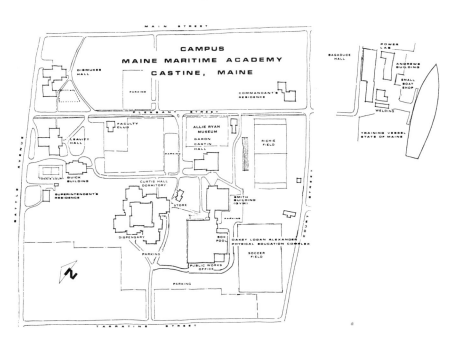

MISSION

The mission of the Maine Maritime Academy is to carry on Maine's heritage of the sea by providing for young men and young women, as cadets, USMS, a comprehensive course of instruction and training in a professional, intellectual, and military environment that will qualify them for leadership as officers in the U.S. Merchant Marine and the U.S. Naval Reserve and as responsible citizens in society.

THE PROGRAM

Maine Maritime Academy is a four-year resident college, offering a program for the young man or young woman interested in a maritime-oriented career. All students are required to live in quarters assigned by the Commandant of Midshipmen.

The Merchant Marine Licensing Program is designed to qualify students for a Bachelor of Science degree in either marine engineering or nautical science, a Merchant Marine officer's license, and a commission as an ensign in the U.S. Naval Reserve or Coast Guard Reserve. Three areas of emphasis—the academic, practical training, and officer development requirements—are involved in a satisfactory completion of this program.

In addition to the above, the curriculum provides for a number of electives that satisfy requirements of minor concentrations in business, transportation, humanities, natural science, social science, oceanography, ocean engineering, engineering science, and marine industrial management.

The Academy receives budgetary support from the State of Maine and the federal government.

HISTORY

Maine Maritime Academy was founded in 1941, by an act of the Ninetieth Legislature. The act followed several years of discussion by leading officials and other personalities in the state urging the establishment of an institution dedicated to maritime training. The first class of twenty-eight assembled in Castine on October 9, 1941, lodged temporarily in the Pentagoet Inn. Rear Admiral Douglas E. Dismukes, USN (Ret) was named as the first Superintendent.

In July 1942, the Academy moved permanently to its present location, taking possession of the grounds and buildings of the Eastern State Normal School, which had ceased operation. During the war years the United States Navy and War Shipping Administration contributed generously to the operation of the school. Midshipmen during those years were Midshipmen, United States Naval Reserve and the staff and faculty, for the most part, were Naval personnel assigned on active duty to the Academy. The first class of twenty-seven men graduated on May 29, 1943, and entered upon active duty in several branches of the armed forces and the merchant marine.

Following World War II, the accelerated curriculum was expanded from eighteen months to the original concept of a three-year program, with graduates being awarded the Bachelor of Marine Science degree. With the outbreak of the Korean War, graduates once again found themselves in demand for positions with the Navy and transport components, bringing their practical know-how in ship navigation, engineering, and naval science to play in facing the responsibilities of service at sea during an emergency.

Peacetime brought further expansion of the Academy program, including a change to a four-year college program that earned a Bachelor of Science degree. The Academy set forth to improve its academic standing, which was achieved with the granting of accreditation, in 1971, by the New England Association of Schools and Colleges. The need for additional officers during the Vietnam emergency led to a further expansion of student enrollment, and a major construction program, which has resulted in today's modern and attractive campus.

LOCATION

Castine, the home of the Maine Maritime Academy, is a historic

Aerial View of Upper Campus.

landmark situated on the shores of Penobscot Bay in Hancock County, Maine. It lies at the end of a peninsula on the east shore of the Penobscot approximately thirty-eight miles south of Bangor, the nearest urban center, and closest commercial airport.

APPLICATION AND ADMISSION

Approximately 185 freshmen will be admitted each year. Final selection rests with the Admissions Committee, which carefully reviews all information contained on the application and in the applicant's file. To be accepted for admission at the Maine Maritime Academy as a cadet, USMS, an applicant must be a high school graduate or have an equivalency certificate; must be a citizen of the United States; must not have reached his twenty-third birthday upon the day of enrollment as a cadet, and must meet the standards established by the U.S. Coast Guard for an original license as a ship's officer. These requirements are as follows:

Height: No limitation prescribed.
Weight: Proportional to height and age.
Vision: For an original license in the nautical science course, the candidate must have uncorrected vision of not less than 20/100 in both eyes, correctable to at least 20/20 in one eye and 20/40 in the other.

For an original license in the marine engineering course, the candidate must have an uncorrected vision of not less than 20/100 in both eyes, correctable to at least 20/30 in one eye and 20/50 in the other.

All applicants must be able to pass the "Stillings" test or the "Williams Lantern" test for color sense. Color blindness disqualifies an applicant for cadet, USMS.

The physical status of candidates will be determined in the following manner. Medical forms to be used to record the results of the physical examination will be sent to the candidate. The examination will be conducted by the candidate's family physician, and the completed medical forms will be returned by the physician to the Director of Admissions.

Candidates who pass this physical examination and meet all other requirements will be judged provisionally accepted. They may, however, be required at a later date to complete another physical examination. Should this subsequent examination show physical deficiencies below the standards established for an original license, the applicant may be disenrolled for medical reasons.

Applicants who have applied for the NROTC program, or for admission to a Service Academy may ask the Department of Defense Medical Review Board to forward a copy of their complete physical examination report to the Maine Maritime Academy.

The Academy is able to offer admission as a cadet to some candidates who may not meet the physical, age, and citizenship requirements. Students who do not meet such requirements will not be eligible to receive a federal subsidy. Upon enrollment, a cadet will be expected to follow all phases of the Academy program, including cruises, summer program, and classroom activity.

Scholastic Preparation

Minimum preparation for pursuing the course of study in the marine license program at the Maine Maritime Academy must include all of the following:

English	4 years
Algebra I	1 year
Algebra II	1 year
Plane geometry	1 year
Physics or chemistry	1 year

The College Entrance Examination Board Scholastic Aptitude Test (SAT) is required of all candidates for admission. The SAT exam should be taken as early as possible in the senior year, but not later than January.

Application Procedures

Applications for admission to the Maine Maritime Academy should be made to the Director of Admissions, Maine Maritime Academy, as early as possible in the senior year of the secondary school course, and not later than April 15 for out-of-state students and by June 15 for in-state students. A nonrefundable application fee of $15 is required with each application.

The following items must be forwarded to the Director of Admissions as early as possible after application has been made:

A transcript of the secondary school record through the first marking period of the senior year.

A letter of recommendation from the principal or guidance counselor of the last secondary school attended.

Three certified copies of the birth certificate with raised seal. These certificates may be obtained from the town/city clerk where you were born or from the Division of Vital Statistics in the state where the birth is recorded.

Aptitude for Merchant Marine Licensing Program

An applicant's file normally includes information indicating aptitude for Merchant Marine officer training. Aptitude in this respect embodies character, morals, extracurricular activities, and interest in the science of the sea.

A personal interview is desirable, and will be arranged if considered to be essential in the process of selection.

Advanced Placement

The Academy may grant credit for an introductory course in an academic field to an enrolled student who presents evidence of competence in that field by completing the appropriate Advanced Placement Examination of the College Entrance Examination Board with a score of four or better. A score of three will be examined for full credit on an individual basis. Courses for which credit may be granted are mathematics, English, chemistry, and physics.

Transfer Students

Candidates for admission from other four-year college programs or two-year junior college and vocational-technical programs will be considered for standing on the quality of their credentials. Such candidates are required to present an official transcript of work completed at the institutions attended. The amount of transfer credit granted will depend on grades earned and similarity to the curriculum offered at the Academy.

Transfer students should note that it is not ordinarily possible to satisfy the nautical science or marine engineering program requirements in less than three years at the Academy.

Applicants for transfer should submit the following items:

1. Application for admission, a letter indicating interest in transfer, and a nonrefundable $15 fee;
2. A complete certified transcript of all grades received at institutions of higher education;
3. A letter of recommendation from the Dean of Students of the last institution attended; and
4. A complete record of secondary school work, including the most recent college board scores.

The applicant for transfer will be required to have an interview at the Academy with the Director of Admissions and the Academic Dean. Visits should be planned during the normal working week.

FINANCIAL INFORMATION*
Schedule of Fees (1977–78)

	In-State	Out-of-State
Tuition	$1075	$2125
Medical fee	75	75
Room	550	550
Board	1050	1050
Cruise fee (freshmen, sophomores, juniors)	50	50
Graduation fee (seniors)	10	
Uniform and book deposit:		
Freshmen, (August 1)	600	
All students per semester after entrance	100	
Room security	50	

* Subject to change

The uniform and book deposit is a credit against purchases in the Academy book and uniform store. Charges incurred by a student in excess of the deposit will be payable on the subsequent bill.

It should be noted that although the above schedule of charges is projected over a four-year period, fees and other charges are subject to such changes as may be ordered by the Board of Trustees during the period of the student's enrollment.

Students who are awarded the federal subsidy to help defray costs of subsistence, books, and uniforms should currently anticipate $600 per year paid in monthly installments and credited to their accounts. This payment is usually in arrears and is not credited in advance.

Schedule of Payments

	In-State	Out-of-State
Freshman year		
May 1 (deposit)	$ 50.00	$ 50.00
August 1 (book and uniform deposit)	600.00	600.00
August 26 (reporting date)	1387.50	1837.50
January 1	1362.50	1812.50
Sophomore year		
July 1	1387.50	1837.50
January 1	1362.50	1812.50
Junior year		
July 1	1387.50	1837.50
January	1362.50	1812.50
Senior year		
July 1	1387.50	1837.50
January 1	1322.50	1772.50

After the first semester of freshman year, and each semester thereafter, the balance of store and room deposits is credited to the student's account and the required deposit of $150.00 is added to the bill due. Payment figures stated above do not include deposit amounts.

In assessing payments due, a credit will be given for student financial aid awarded, prorated on a semester basis. The balance due must be paid within thirty days of due date or a late payment fee of $10 will be added to the account. Overdue accounts may accrue interest beyond the thirty-day grace period at the rate of one per cent per month on the unpaid balance. Students ordinarily will not be permitted to register for classes unless all accounts are current. Subsidy as received is credited to balances due at the time statements are prepared.

Federal Subsidy

With the exception of a limited number of special students, all students attending the Maine Maritime Academy will be enrolled as cadets, USMS. Each fall, the U.S. Maritime Administration assigns a

Aerial view of academy waterfront.

number of subsidy positions to be allocated to members of the new freshman class. Since the number is limited, not all physically qualified cadets, USMS, can expect to receive the subsidy. Allocation of available positions is usually made toward the end of the first semester, and is governed by criteria established by the Academy Superintendent. At present the subsidy accrues at the rate of approximately $50 per month from the date of enrollment, and is credited directly to the individual student's account as it is received from the Maritime Administration.

The Maine Maritime Academy receives a federal grant of $75,000 a year to assist in its operation. The training ship *State of Maine* is loaned and maintained by the federal government for the purpose of training cadets, USMS.

Financial Assistance

Financial aid of various types is available to students at Maine Maritime Academy. Any student who is already enrolled or any candidate who has been accepted for admission may apply for financial aid.

Applications for financial aid for the coming academic year must be completed prior to June 15. Entering students who complete the application procedure prior to April 15 may be notified of awards for the coming year by May 1.

Meeting the financial obligations of a college education is the prime responsibility of the student and family. Where a need for aid is shown, the Academy may assist with loans, scholarships and/or part-time work opportunities as funds are available.

All applicants must have a Parents' Confidential Statement (PCS) and a Financial Aid application on file before any awards may be made. The PCS, which is processed by the College Scholarship Service, Princeton, New Jersey, is available at any high school guidance office.

A financial aid application and the Parents' Confidential Statement or Student's Financial Statement are required for each academic year that the student wishes to receive financial assistance. Applications are available from the Financial Aid Office, Maine Maritime Academy.

STUDENT HEALTH

The Maine Maritime Academy maintains a well equipped medical department and infirmary under the direction of a full-time physician. There is a pharmacist in attendance, as well as a student health officer who lives at the infirmary.

Sick call is held twice daily. In addition, one of the medical officers is always on duty should any condition warrant attention. Many cases are handled in the infirmary, under the care of the Academy physician. More urgent cases are treated at the Castine Community Hospi-

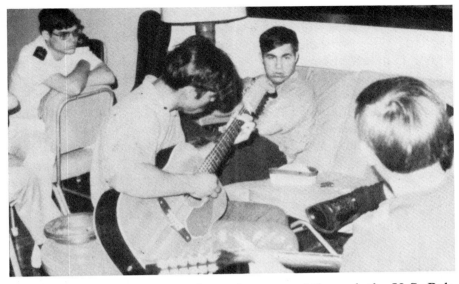

tal, or other hospitals. Dental care is arranged through the U.S. Public Health Service at Portland, Maine.

All students at the Academy are charged an annual medical fee, which provides for their enrollment in a group Blue Cross and Blue Shield program. All students enrolled as cadets, USMS are eligible to receive U.S. Public Health Service treatment at any USPHS facility, free of charge, upon enrollment at the Maine Maritime Academy.

All students are provided complete immunization after entering the Academy and prior to all cruises, as deemed advisable by the medical officer. During training cruises, a qualified physician is aboard the ship.

LEAVE

Leave, with allowances from the government for cadets, USMS, the specific periods of which shall be at the discretion of the Superintendent, shall be as follows:

> If transferred to a hospital, or sick at home, or in the sick bay—not to exceed four months;
> For an emergency due to the serious illness, injury, or death of a very near relative—not to exceed seven days;
> Annual leave—not to exceed thirty days; and
> Christmas and Easter—not to exceed a total of twelve days.

THE REGIMENT OF MIDSHIPMEN

Students entering and attending the Maine Maritime Academy are organized into what is referred to as the Regiment of Midshipmen. This involves all four classes; freshmen (fourth class), sophomores (third class), juniors (second class), and seniors (first class). The Regiment consists of two battalions, four companies, eight platoons plus the ship and service rates, a band, and a drill team.

All students attending the Academy are required to be in a military uniform at all times while at the Academy—similar in design to that of the U.S. Naval officer's uniform.

The administration, control, accountability, and discipline of the Regiment is handled by the midshipmen themselves, with considerable authority and responsibility invested in the senior class—the first classmen under the control, direction, and guidance of the Commandant of Midshipmen and his staff officers.

The Regiment of Midshipmen is under the command of a Regimental Staff headed by the Regimental Commander, who serves as the Regiment's direct contact with the Commandant of Midshipmen and the Assistant Commandant.

The daily routine of the Regiment is governed by the rules and regulations explicitly written in the *Regimental Manual*. Included in the manual is a demerit system.

It is incumbent on each applicant who matriculates at the Academy that he or she understand this aspect of Academy life. More important, each new student should bring to the Regiment a commitment to expend his or her energy and talents on behalf of the goal of bringing the Regiment to even higher standards of excellence.

There is much truth in the statement that when the first class graduates and the new fourth class enters all seems in a state of transition, but in fact the Regiment is constant; it changes only in its excellence of performance.

As the midshipmen and officers alike live up to their responsibilities the Regiment will exhibit a sense of confidence and well-being apparent to all.

An integral part of the Regiment of Midshipmen that lends it much color and prestige is the Academy Band, with the Drill Team and Color Guard.

The Academy Band is a volunteer activity organized and directed by the midshipmen with limited faculty assistance. The Academy Bandmaster is a senior midshipman who has come up through the ranks of the Band.

EXTRACURRICULAR ACTIVITIES

There are many opportunities for students to be involved in extracurricular activities at the Academy. Some of the active student organizations include:

Academic Council
Maine Maritime Academy Yacht Club
World Trade Club
The Maine Maritime Academy Chorus
The Singing Mariners
Midshipman Council
The Amateur Radio Club
Cadet's Radio Station (WMMA)
Scuba Club

The Singing Mariners.

Omega crew stands by for a day of sail.

Student Publications
Karate Club

ACADEMIC POLICIES

Scholastic achievement is determined by the average of the grades received by each student in all subjects. Each course is given a weight proportional to the number of credit hours assigned. The weighted average of all grades will constitute the student's scholastic average for the academic semester.

Grades will be reported as A, B, C, D, and F, where A is superior;

B   good, above average but not superior
C   average
D   low grade, below average but passing
F   failure (no credit received)

Grades will be given quality point values as follows for each credit: A, 4 points; B, 3 points; C, 2 points; D, 1 point, and F, 0 points.

To establish eligibility for receiving the Bachelor of Science degree from the Maine Maritime Academy, each candidate for the degree must present the following evidence:

1. Successful accumulation of 140 semester credit hours (SCH) with a cumulative quality point average of not less than 1.9;
2. Successful accumulation of 12 semester credit hours (included in the 140 SCH listed above) for satisfactory completion of the two training cruises and the cadet shipping program. A third training cruise may be substituted for a cadet shipping program in the event that berths in the cadet shipping program are not available;
3. Successful completion of the license preparation program; and
4. Successful completion of the practical and military requirements.

FRESHMAN ACADEMIC PROGRAM

During the first year the student normally takes basic courses in selected fields that familiarize him or her with various modes of learning and prepare them for the advanced nature of study ultimately required in their major field. The curriculum is designed to allow a student to defer the choice of nautical science or marine engineering as a major until the end of the freshman year. A limited number of transfers from one program to another may be possible at a later time.

PRACTICAL TRAINING PROGRAMS

In addition to the Academic programs that students pursue at Maine Maritime Academy, there are practical training programs that are also part of the overall requirements for graduation. The practical training programs consist of the following:

Ship laboratory and watch standing
Annual cruise of the Academy training ship
Cadet shipping cruise

Ship Laboratory and Watch Standing

During noncruise periods, students are expected to satisfy student watch standing requirements of the Academy. Student watches are maintained in Castine at all times except when the training ship is on annual cruise. To minimize classroom absences, watches are kept to those essential to provide reasonable security for the Academy and training for students.

Students are required to participate in ship's laboratory in maintaining the training vessel and in gaining practical experience.

While absences from classes are occasioned by the demands of watch standing and ship's laboratory, it is, nevertheless, the respon-

T.S. State of Maine in Helsinki Harbor. Photograph taken from U.S. Embassy grounds.

sibility of students to keep up with classroom work and progress through self-study and consultation with their instructors. Out-of-class assignments in each course are to be performed by students who are engaged in these duties.

Training Ship Cruise

Training cruises are conducted annually during May and June, commencing shortly after final examinations. Students are required to take the training cruise after their freshman and junior academic years. Successful completion of training cruises, including a sea project for each cruise, is required for graduation. Four semester credit hours are awarded for each cruise. The cruise is designed to bring into practical focus the various skills and training required of a third mate and third assistant engineer through watch standing and experience in the operation and maintenance of the ship and all its appurtenances.

Lifeboat inspection and drill.

Academy training cruises normally have an extensive itinerary including European or Caribbean and South American ports, and provide the student with an opportunity, not only to gain shipboard training experience, but also to observe at first hand foreign countries and their people. The cruises provide social and cultural growth opportunities that are not available to students in normal college programs.

Cadet Shipping Program

Upon completion of the sophomore year, students in the nautical science and marine engineering programs may be assigned aboard U.S.-flag merchant ships as Cadets for further familiarization in shipboard procedures, and training in the responsibilities they will be given as licensed third mates or third assistant engineers. A minimum

At the helm, *T.S. State of Maine.*

Chief engineer's project during ship's maintenance.

Plotting a course in the ship's chart room.

of sixty days is required for this training, which is credited toward the sea service requirement for an original license in the Merchant Marine. A sea project is required of each student, and four semester credit hours are awarded for successful completion of the project and cruise. Successful completion of the cadet shipping program or a third cruise on the training ship is a requirement for graduation.

Aside from the practical experience gained, students may have the opportunity to visit ports of call both in the United States and foreign countries. Some students, depending on their ship assignments, have an opportunity to circumnavigate the globe in this program.

ACADEMIC PROGRAMS AND MARITIME DEPARTMENTS

The degree of Bachelor of Science is conferred upon graduates of the Academy with majors in nautical science, for deck officers, and marine engineering, for engineer officers. Each candidate for the Bachelor's degree must accumulate 140 semester credit hours, and must take certain required courses listed in the curriculums below for the two major programs. The student is required to take five elective courses commencing in the fall of his or her sophomore year, thereby allowing them to pursue fields of specific interest to the individual.

All freshman take the same courses; these are principally in the area of the arts and sciences with modest introductions into deck and engineering subjects. At the end of the freshman year, the student elects his or her choice of major program, i.e., nautical science or marine engineering. Commencing in the sophomore year, therefore, they pursue the studies listed below in each of these areas of major concentration.

Department of Nautical Science

The program offered by this department gives the professional courses required to qualify the deck student to receive a degree and, after passing the required U.S. Coast Guard examination, a federal license in the Merchant Marine. This curriculum is designed to familiarize the deck student with all phases of navigation and piloting, rules of the road, rules and regulations, deck seamanship, cargo handling and stowage, visual signaling, ship handling, and management. Electives are offered by the Department.

Department of Marine Engineering

The Engineering Department offers courses that pertain directly to design, operation, and maintenance of marine power plants. These courses are required to qualify the engineering student to receive a degree and, after passing the required U.S. Coast Guard examination, a federal license in the Merchant Marine. The department not only offers a thorough training in the fundamentals of engineering, but also

262

coordinates theory and practice by relating classroom studies to the student's practical experience aboard ship and in the machine shop and steam and nuclear propulsion laboratories. Electives are offered by the Department.

Department of Naval Science

Our national defense requirements demand the backing of an adequate Merchant Marine manned by well trained officers who are able to operate efficiently with the Navy in time of war or national emergency. Accordingly, although the Navy does not desire the operation of any maritime academy for the sole purpose of training naval officers, it does have a distinct interest in the education afforded students at these academies. All students are required by the Academy to enroll in the naval science courses offered by the Department of Naval Science. The courses taught are those prescribed by the Chief of Naval Education and Training, and are the same courses taught nationwide as a part of the ROTC program. The courses are taught to provide the student with the primary knowledge and skills required of a junior naval officer, to familiarize the student with the relationship between the Navy and the Merchant Marine, and to qualify the student for a commission in the Naval Reserve. All naval science courses are taught by active-duty officers assigned to the NROTC unit, Marine Maritime academy.

Students hook up wiring on a tanker simulator.

Minor Concentration

All students are required to take five elective courses, which they may elect in one area of academic discipline to constitute a minor when combined with a required core course in the same area. In the event that a student does not wish to have the concentration in depth afforded by a minor program, he or she may choose electives in different areas of academic disciplines. To complete successfully a minor program, students are expected to achieve a grade of at least C in their minor courses. Students commence the elective program in the fall of their sophomore year. They may choose from these minor programs:

Business/Transportation
Engineering Science
Humanities
Marine Industrial Management
Natural Science
Ocean Engineering
Oceanography
Social Science

License Preparation

The purpose of the license preparation program is to provide for seniors a comprehensive review of their professional courses as a part of their professional program of studies. It is an integral part of the overall Academy program and, therefore, participation in, and successful passing of, the program is a requirement for graduation from the Academy. No credit is given for the program since credit for this material will have been previously awarded when covered in specific courses in the program.

Capitulation of Credits:	*Semester Credit Hours*
Academic credits in program:	
nautical science	142.5
marine engineering	146.5
Two cruises and Cadet Shipping	12
Required for graduation	140

ATHLETIC PROGRAMS

The varsity athletic program is designed to give the highly competitive and highly skilled athlete an opportunity to excel in a chosen sport. The Academy sponsors six varsity teams: football, cross-country, golf, soccer, wrestling, and indoor track.

PLACEMENT OFFICE

A placement service is maintained by the Marine Maritime Academy to assist members of each graduating class to obtain employment and also to aid alumni seeking new positions. This service is not restricted to maritime industry. The office maintains contacts with many of the major industrial corporations, and many graduates are serving with distinction in positions of responsibility as engineers, managers, sales and service representatives, and in a wide range of capacities under civil service. A number of recent graduates have continued their education in graduate programs in such fields as business management,

Graduation day.

education, and engineering.

Each year a number of graduates select the military as a professional career. Commissions as ensign either in the U.S. Navy or the U.S. Coast Guard can be available to the qualified graduate.

Among the companies who have hired our graduates in recent years are: Bethlehem Steel Company; Ford Motor Company; Bath Iron Works; Central Maine Power Company; Marine Yankee Atomic Power Company; Worthington Corporation; General Dynamics, Electric Boat Division; General Electric Company; and numerous U.S. maritime companies, both oceangoing and in the Great Lakes.

The employment opportunities for the Marine Maritime Academy graduate are many and varied, and he or she is amply trained and prepared to undertake his or her choice of profession.

APPLICATION INFORMATION

For further information concerning the Maine Maritime Academy write to: Admissions, Maine Maritime Academy, Castine, Maine 04421.

Great Lakes Maritime Academy

The Great Lakes Maritime Academy, a division of Northwestern Michigan College, prepares young people to serve as officers aboard Great Lakes ships. Included in the three-year program is extensive training in seamanship and engine mechanics, while sailing on the Great Lakes.

The program is unique among state maritime academies in three ways.

First, it is the only one that operates as a division of a community college offering the degree of Associate of Applied Science and qualifying all successful cadets to be examined under U.S. Coast Guard regulations for either a first class pilot license (Great Lakes) or third assistant engineer's license.

Second, cadets are part of the student body at NMC, live in the residence halls with other students, and are not required to follow a rigid military schedule. They are required to carry out their academic responsibilities effectively and to demonstrate the ability of self-discipline and leadership required in the exercise of good seamanship. Cadets wear uniforms in maritime classes when designated.

Third, the program is the only state maritime academy now operated in the United States on fresh water.

All cadets are required to abide by the *Rules and Regulations Governing Cadet Discipline,* which may be revised as needed. The Great Lakes Maritime Academy reserves the right to revise the program in accordance with the needs of industry and the requirements of governmental agencies.

The curriculum includes subjects in four major areas:

1. Basic arts and sciences, including English, mathematics, physics, history, government and psychology.
2. A number of maritime-allied technical courses, such as electronics (with special emphasis on marine usages of electronics), an introduction to electronic data processing, and engineering drawing. Students who elect to study and train toward the deck officer objective take astronomy and meteorology. Those who choose the engine officer option take machine shop practice, metal-working, and electrical engineering.
3. Maritime courses, which include ship construction, ship stability, labor relations, and shipboard medical practices. Deck cadets take cargo stowage, nautical rules of the road, maritime law, ship management, and several courses in navigation. Engine cadets have an extensive program in steam and diesel engineering.
4. Sailing requirements include on-the-job experience and professional correspondence assignments on board ship. There are also academic courses to complete at this time.

The Academy program is approved by the United States Maritime Administration, the United States Coast Guard, and the Michigan Department of Education, but is not a naval science (U.S. Navy-sponsored schooling) maritime academy.

ENTRANCE REQUIREMENTS

Admission to the Academy is based on the same standards required of all graduation candidates entering Northwestern Michigan College and similar standards used for admission to other state maritime academies, with the exception that cadets at the Great Lakes Maritime Academy are not required to be unmarried at the time of admission nor to remain single during the program.

Students who have completed some of the required courses elsewhere will receive credit where applicable.

To qualify for United States Maritime Service, each applicant must:

1. Be a citizen of the United States.
2. Meet all U.S. Coast Guard requirements, including physical for issuance of original license for merchant marine officer.
3. Be no less than seventeen years of age.
4. Have a secondary school diploma or equivalent, satisfactory for admission as an undergraduate. To be satisfactory, secondary schooling must include algebra and science, including chemistry, physics, or both, and the applicant must have an academic grade point average of 2.0 or higher, and a minimum of 40 percent ACT composite.

In addition to academic criteria, character references, work experience, and general accomplishments in high school are considered in reviewing a candidate's application.

Students admitted to the Great Lakes Maritime Academy, who have applied for, and been enrolled in, the U.S. Maritime Service will be entitled to receive a uniform, textbook, and subsistence allowance, provided that Congress has appropriated necessary funds for this purpose.

This subsistence allowance is approximately $50.00 per month, and is available to only fifty cadets who are admitted, enrolled, and in good standing thru the fall term. Eligible cadets will begin receiving their allowance checks in January; payments are retroactive to August.

OTHER REQUIREMENTS

Maritime cadets are required to maintain a 2.0 or above in *all* maritime subjects. Failure to do so will result in the cadet being dropped from the maritime program until the course is repeated and a grade of 2.0 or better is achieved. Readmission to the maritime program is by appeal in writing to the Director of Admissions and the Division Director.

Maritime cadets are expected to maintain continuous enrollment at Northwestern Michigan College (GLMA) for the duration of the program.

GRADUATION REQUIREMENTS

In addition to the general graduation requirements of the college, academy cadets shall complete the prescribed academy program successfully; pass all professional courses with a 2.0 or better grade ; and meet a minimum of twenty-four months participation in the Great Lakes Maritime Academy program (this applies to transfer students only).

THE CAMPUS

Among the distinguishing features of NMC's main campus are tall pine trees, spaciousness, convenient location, and buildings that blend well with the natural surroundings. The 146-acre wooded campus is near Grand Traverse Bay, within walking distance or a short drive of the business and recreational areas of Traverse City.

Major expansion of NMC's facilities began in 1957, when the school's first building (an all-purpose structure that housed classrooms, library, and cafeteria) was enlarged to provide additional classrooms and space needed for administrative and faculty offices. Situated in the center of the campus, it is now the College Administration Building.

In 1960–61 the Mark Osterlin Library was constructed. Its resources in books, periodicals, and microfilm collections have grown steadily, and the building is often called "the heart of the campus."

Next came a Science Building, a residence hall (since named West Hall) and a Student Center, all completed in 1963. The Student Center, with its dining facilities, lounges, and meeting rooms, is the "Social hub of the campus."

In 1965 the voters of Grand Traverse County approved the necessary tax levy that initiated a five-year expansion program. A wing was added to the Science Building, more than doubling its original size. A second residence hall (East Hall) was constructed. A cherry-processing plant, located on eight acres of waterfront property a mile from the main campus, was purchased and partially remodeled to become the NMC Technical Center, housing classrooms and laboratories for a number of vocational and technical programs.

The Nick Rajkovich Physical Education Center, completed in 1969, provides a campus headquarters for physical education and health classes, intramural sports, student convocations, and community programs.

A Fine Arts Building was completed in 1971, and in that year the College also improved and expanded facilities for its Career Pilot Training program and its Great Lakes Maritime Academy. T-hangers, an operations room, and space for ground trainer instruction was centralized at Traverse City's Cherry Capital Airport. Along the waterfront at NMC's Technical Center, a new dock and marine basin were constructed to accommodate Maritime Academy training ships and auxiliary craft.

A year later, as part of another five-year expansion program initiated again by an approving vote for a tax levy from Grand Traverse County residents, other buildings in the spacious complex of the NMC Technical Center were remodeled to further expand the College's technical training facilities. Simultaneously, student housing accommodations were increased on the main campus by the construction of an apartment complex.

Currently a new college academic building, the Health and Education center, is being constructed on the main campus. Supported by local village and state funds, the 27,000 square foot structure will be completed in the fall of 1976.

THE CURRICULUM

The recommended sequence that follows lists the coursework included in the Great Lakes Maritime Academy curriculum:

DECK OFFICER OBJECTIVE

First Year

Summer	Credits
Maritime Discipline I	6
College Arithmetic (if required)	4
	10

Fall	
Freshman Communications	4
Maritime Discipline II	3
Technical Mathematics I	4
Chemistry	3
Chemistry Lab	1
	15

Winter	
Freshman Communications	4
Maritime Discipline III	3
Marine Electrical Theory	4
Navigation I	4
Technical Mathematics II	4
	19

Spring	
Freshman Communications	4
Communications	2
Nautical Rules of the Road	2
Technical Mathematics II	4
Meteorology	3
Meteorology Lab	1
	16

Second Year

Summer

History of Western Civilization	4
D Sea Project	9
	13

Fall

Navigation II	4
Electronic Aids to Navigation	4
Navigation Electronics Lab	2
Astronomy	3
Astronomy Lab	1
	14

Winter

Computer Programming	4
Maritime Law	3
Ship Stability	3
Shipboard Medical Practice	3
Blueprint Reading	3
	16

Spring

Ship Construction	2
Labor Relations & Ships Management	4
Applied Psychology	3
Physics	3
Physics Lab	1
Elective	4
	17

Third Year

Summer

Survey of American Government	5
D Sea Project	9
	14

Fall

D Sea Project	9

Winter

Navigation III	4
History of Great Lakes Shipping	2
Cargo Stowage	3
Elective	4
	13

Spring
    License Preparation   8

ENGINE OFFICER OBJECTIVE

First Year

Summer	*Credits*
Maritime Discipline I	6
College Arithmetic (if required)	4
	10

Fall	
Freshman Communications	4
Maritime Discipline II	3
Technical Mathematics I	4
Chemistry	3
Chemistry Lab	1
	15

Winter	
Freshman Communications	4
Steam Engineering I	4
Steam Engineering I Lab	2
Technical Mathematics II	4
	14

Spring	
Freshman Communications	4
Auxiliary Equipment	4
Auxiliary Lab	2
Technical Mathematics III	4
Blueprint Reading	3
	17

Second Year

Summer	
History of Western Civilization	4
E Sea Project	9
	13

Fall	
Ship Stability (E)	3
Marine Electrical Theory (E)	4
Diesel Engineering I	4
Diesel Engineering I Lab	2
Applied Psychology	3
	16

Winter

Marine Electrical Engineering I	3
Marine Electrical Engineering I Lab	1
Diesel Engineering II	4
Diesel Engineering II Lab	2
Welding I	4
	14

Spring

Steam Engineering II	4
Steam Engineering II Lab	2
Marine Electrical Engineering II	3
Marine Electrical Engineering II Lab	1
Labor Relations & Ships Management	4
	14

Third Year

Summer

Survey of American Government	5
E Sea Project	9
	14

Fall

E Sea Project	9

Winter

Machine Shop	4
Refrigeration	3
History of Great Lakes Shipping	2
Shipboard of Medical Practice	3
Physics	3
Physics Lab	1
	16

Spring

Maritime Discipline III (E)	3
License Preparation	8
	11

NORTHWESTERN MICHIGAN COLLEGE

In its 25 years Northwestern Michigan College has grown from 65 students to over 2400, from a faculty of 5 to more than 100, from borrowed classrooms at Traverse City's municipal airport to well-equipped buildings on two campuses valued at $8.5 million, and from a handful of courses to a number that is close to 500 and constantly growing.

NMC was Michigan's first community college. Now, throughout

the entire state there is a sizeable network of tax-assisted, two year public colleges that provide educational services described as invaluable in accommodating today's college populations.

Many NMC students are from Grand Traverse County. Others come from all sections of Michigan, many other states, and several foreign countries. Their educational needs and interests are varied, and NMC responds by constantly updating, innovating, and initiating a wide range of studies.

Throughout its twenty-five years NMC has been an influencing factor in the educational, cultural, and economic growth of the Grand Traverse region, and is a thriving example of a community's achievements. Its founding in 1951 was a community effort, and its growth is due in large measure to the continuing support it receives from the area it serves.

Entrance Requirements

Applicants to Northwestern Michigan College are expected to be in good health, of good character, and possess a high school diploma or its equivalent. High school grades, test scores, recommendation by principal or counselor, and special aptitudes are additional criteria taken into account in determining qualification for admission for programs of study that can accommodate only a limited number of applicants each year. Priority consideration is given to qualified applicants who reside within the College's service area.

Enrollment at Northwestern Michigan College is a privilege that carries certain responsibilities with it. The College has established standards of scholarship and conduct for the benefit of all concerned, and it reserves the right to cancel enrollment whenever it judges that a student is unwilling to accept these standards. Northwestern Michigan College's admission practices do not discriminate on the basis of race, color, sex, or national origin.

Regular admission is granted to students who show the potential to accomplish work leading to an associate degree or, in the case of certain specialized programs of less than two years' duration, a certificate of achievement. Regular admission is possible in two ways:

1. Applicants with no prior college study must present a completed high school transcript of C average or better, or its equivalent, and be recommended by their high school principal or counselor. Applicants from the NMC primary service area who have below a C average will be admitted on a provisional basis and will receive special academic counseling. The primary service area of the College includes all of Grand Traverse, Benzie, Leelanau, Kalkaska, and Wexford Counties, and most of Antrim County.
2. A student who enrolls at NMC as a nonresident of the district shall continue to be so classified throughout his attendance as a student, unless and until he demonstrates that a district (Grand Traverse County) domicile has been established. Students can-

not apply for reclassification as district residents until domiciled in the district continuously for six months before the first day of classes of the term for which reclassification is sought. During this six-month period students may be enrolled part time for a maximum of eight credit hours per term. Additional criteria used to determine a student's residency status may be found in the *NMC Guidelines* for administration of the residency policy, which can be obtained from the Dean of Student Services.

3. Applicants with prior college study must present a transcript of accredited college work. Transcripts are individually evaluated to determine the admissibility, but a C average is expected.

Special admission is granted to students whose status is defined in one of the following three ways:

1. Part-time students are those who enroll for one or two courses and are not seeking the associate degree. When these students show evidence of ability to do satisfactory college-level work, they may apply for regular admission.
2. Guest students are those students who enroll for one term to take selected courses for transfer to another college.
3. Noncredit (audit) students are those who do not wish to receive credit for courses taken. These students are required to pay the same tuition and fees as credit students, but they do not receive grades. Enrollment is limited to courses where space is available and class procedure is amenable to auditing.

Credit by Examination

Northwestern Michigan College may grant credit or allow advanced placement on the basis of satisfactory scores on college-level examination program tests (CLEP). The College recognizes only the Subject and not the General tests. Full details are available in the Counseling Center.

Admission of Students from Out of State

Admission for out-of-state students generally requires a B average. However, available space will be a determining factor for admission to many programs of study. Sons and daughters of alumni are considered the same as Michigan residents for admission purposes only.

Admission of Students from Other Countries

Northwestern Michigan College welcomes applicants from other nations who show promise of profiting from an educational experience here. The College requires proficiency in English as evidenced by the results achieved in tests such as the TOEFL tests (Test of English as a Foreign Language) or by other certification of proficiency. Students from other countries are required to purchase student health insurance coverage if not otherwise insured.

They must have the equivalent of a U.S. high school education with an approximate grade average of B. Since the College does not have financial aid available for students from other nations, the applicant must clearly demonstrate his ability to finance his education in the U.S. by completing a financial statement. Prospective applicants should write to the Director of Admissions for complete application procedures at least six months prior to the term of enrollment.

Testing Requirements

New students are required to submit the results of the American College Test (ACT) before registration for classes. All beginning full-time freshmen and part-time students pursuing associate degrees or planning to enroll in English or mathematics classes must take the ACT. Further, it is recommended that all students take this test for state scholarship consideration and for the information about career interests that is provided with the test results.

The ACT results are used to aid the student and the counseling staff in planning programs and choosing appropriate courses.

Summer Counseling and Orientation Program

All entering freshmen are required to participate in a Summer Orientation Program on campus, at which time they will become acquainted with the campus, meet members of the NMC faculty and staff, plan their studies, and register for the fall term. Each admitted student then validates his registration by securing a student identification card and by paying tuition and fees. An extended payment plan is available and is descirbed later in this chapter.

A special orientation and registration program is also required for all new freshmen entering the winter or spring term.

Counseling and early registration for winter, spring, and summer terms will take place during the final weeks of the preceeding terms for currently attending students.

Application Procedures

Application forms are available in high schools or may be obtained from the Admissions Office at the College. Students are encouraged to apply during the first semester of their senior year in high school. The procedures to follow are these:

1. Fill out the student portion of the application form and attach the nonrefundable $10.00 processing fee, in the form of a check or money order payable to Northwestern Michigan College.
2. Turn in the application to your principal, counselor, or registrar,

who will complete the remainder of the form including a recommendation statement, enclose a transcript of your record and forward it to the College.

3. Request a transcript for all posthigh school education from the Registrar of each institution previously attended.
4. Take the required American College Test (ACT) and request that the test results be sent to Northwestern Michigan College.
5. All new students are required to complete a health form for use in the Student Health Service Office. Some programs, including Nursing and the Great Lakes Maritime Academy, require a health form completed by a physician.

Veterans and Veterans' Children

Northwestern Michigan College has been recognized by the Veterans Administration as qualified and equipped to furnish education at the collegiate level under the provision of Public Law 89-358, "Veterans Readjustment Act of 1966." Veterans are urged to contact the Veterans Affairs Office well in advance so that forms are ready when admission and enrollment is completed. A transcript evaluation of any previous college training is required.

NMC has published a list of standards of progress for veterans, which is available in the Veterans Affairs Office. Veterans enrolling under PL 89-358 are required to pay their tuition and fees and buy their books and supplies just as civilian students do.

War Orphans

If a student is the son or daughter of a deceased or totally disabled veteran, he or she may be eligible for government assistance to attend college. Students should contact the Veterans Affairs Office for information.

Learning and Working Flexibility

It is the intent of NMC to make it convenient for all students to further their education. To accomplish this, the College offers flexible programming so that students may develop work and study schedules tailored to their individual needs. Among the options are these:

1. Students may work full time and take courses part time.
2. Students may study full time and work part time.
3. High school students, with permission of their school, may take NMC courses concurrently.
4. Students attending other colleges may enroll as guest students at NMC during any term.
5. Students may enroll at NMC for a term, work during the next term, then re-enroll, or arrange similar schedules in combinations of terms.

6. Students may enroll continuously year-around to complete their program of studies at a rapid pace.

Housing

There are two college-operated residence halls and an apartment complex on the main campus of NMC. West Hall provides living accommodations for 160 to 200 men. East Hall provides living accommodations for 240 to 300 women. Both residence halls contain recreational facilities, lounges, meeting rooms, study rooms, and laundry equipment for student use. The residence hall student will be in a position to derive maximum returns from College activities, since learning extends beyond the classroom. Opportunities for personal and social growth through group living should add immeasurably to the total collegiate education.

NMC requires all unmarried freshmen who do not live with their parents or relatives to reside in College housing. If space is not available in the housing facilities the Housing office maintains a list of possibly available accommodations in the community. Generally, off-campus housing compares favorably in cost with the residence halls.

Information and application blanks for campus housing reservations will be sent to students after their letters of acceptance to NMC (usually in April for the following fall term). Applications for housing should be returned to the College as soon as possible, and must be accompanied by a $50 deposit.

The apartment complex, completed in 1973, has one-and two-bedroom apartments for 138 male and female full-time sophomores. Married students may live in the complex if they choose. Each apartment is carpeted, furnished, and has storage facilities. Bed linens and cooking utensils are not furnished. Each of the three buildings in the complex has laundry facilities for use by its residents. Apartment residents are required to sign a contract and submit a $50 deposit.

For information, address inquiries to: Co-ordinator of Housing, Northwestern Michigan College.

Student Activities

At Northwestern Michigan College co-ed and extracurricular programs that stimulate and reflect students' interests are considered an important part of college life, second only to studies themselves. Outside the classroom, students are encouraged to develop, and participate in, programs of interest to them to complement classroom instruction, develop social interaction, and provide profitable use of leisure time.

Clubs and organizations for students with special interests—a variety of cultural programs, seminars and lectures on contemporary subjects led by area, state, and national resource persons, student and faculty participation in the arts, the student-operated College radio station, WNMC—are all part of the student activities scene.

In the intramural and sport club area the College engages in three kinds of competition. Intramurals: flag football, basketball, and slow pitch softball. Sports Club: Tae Kwon Do, judo, power volleyball, bowling, and skiing. Athletic Sports Clubs: baseball, tennis, golf, and cross-country. Neighboring recreational areas make their facilities available to NMC students. These, plus the on-campus facilities of the Nick Rajkovich Physical Education Center, enable men and women students to participate in a rapidly expanding intramural and sports club program.

Strictly for fun, there are NMC mixers, dances, picnics, and ski-outs held regularly throughout the year and usually sponsored by College clubs and the Student Council.

Student Health Service

All students registered in regular credit courses at NMC are eligible for services at the Student Health Center upon payment of the health service fee ($3.00 each term). A pre-entrance health evaluation form must be on file, and a validated identification card shown before services may be rendered. The services consist of the treatment of ordinary illnesses, injuries, and other health problems.

In case of serious illness or injuries that require hospitalization, the use of hospital facilities (emergency room, X-rays, etc.), or a visit to the office of a private physician, the cost will be born by the student through his insurance or by direct payment to the physician or hospital.

Student Health and Accident Insurance

Students are *not* automatically covered by college insurance policies, but a low-cost health insurance policy for students is available through the college. Obtain information and claim forms at Health Service in the Health and General Education Building. If the student is covered by his own or his parent's insurance policy (Blue Cross, Prudential, etc.), he should follow the claim procedures suggested by the company.

Financial Information

Tuition
In-district
(Grand Traverse County residents

$11.50 per credit
hour per term

In-service area
(residents of Antrim, Benzie, Kalkaska,

$19.00 per credit
hour per term

Leelanau, and Wexford
Counties)
In-state $21.00 per credit
(residents of other   hour per term
Michigan counties)
Out-of-state $23.00 per credit
                                     hour per term

General fees
$2.00 per credit hour per term
   (These fees apply to all students, and cover student publications, Student Center, extracurricular activities, and other services connected with College classes.)

Other fees
Application fee $10.00
Health service fee (per term)   3.00
Late registration fee (after $10.00
close of announced
registration periods)

Residence halls
   Room and board, $1,280 per year divided as follows:
   Fall term   (12 weeks) - $450.00
   Winter term (11 weeks) - $415.00
   Spring term (11 weeks) - $415.00

   A $10.00 discount will be given per term if students pay the full amount at beginning of each term.
   Room deposits, $50, must be submitted with residence hall applications. Cost may be paid in full or by the term according to the payment schedule on the reservation cards.

Apartments
Fall term $305.00, 2-bedroom; $320.00, 1-bedroom
Winter term $305.00, 2-bedroom; $320.00, 1-bedroom
Spring term $305.00, 2-bedroom; $320.00, 1-bedroom
   A $10.00 discount will be given per term if student pays the full amount at beginning of each term.

Extended Payment Plan.   Students may, without extra charge, pay tuition and fees in two installments each term. Under the extended payment plan, students may pay one-half of the total tuition and fees at the time of registration. (In case of part-time students whose total tuition and fees are less than $100, full payment or a minimum of $50 will be required at the time of registration.) The balance will be due midway through the term.

Scholarships and Financial Aids.   NMC believes that paying for college is a shared responsibility Educational costs are to be shared among the student, who may have savings from summer employment and from loans or jobs during the academic year, parents, who should make a reasonable contribution from their income and assets, and the college or scholarship agency, which may be able to offer financial aid.

The goal of the college in promoting a financial assistance program is to remove economic barriers to postsecondary education among the able people of all classes of our society. To accomplish this, NMC presents a number of alternative resources designed to help the individual student meet his or her needs.

Various scholarships, grants, grants-in-aid, and loans are available to insure maximum opportunity for aid. Northwestern Michigan College's financial aids range in value from scholarships that pay all tuition and fees to awards that cover roughly one-third of that expense.

Scholarships are administered by the College Scholarship Committee, and are awarded primarily on the basis of scholastic ability and financial need.

Grants-in-aid are also administered by the College Scholarship Committee, and are awarded to students primarily on the basis of financial need and stipulations stipulated by the donor.

Awards are given annually to students who have achieved outstanding success in particular fields of study, or who have made noteworthy contributions in extracollegiate activities to significantly benefit the College and the student body. Many of these awards are donated by community residents and organizations. All are administered by faculty members who direct the academic discipline or supervise the activity for which the award is made.

Loans are available from a variety of funds made available by individuals and organizations and from the National Direct Student Loan Funds, and generally are made only to students who have demonstrated ability to do satisfactory college work.

Northwestern Michigan College participates in the College Scholarship Service (CSS) of the College Entrance Examination Board. Participants in CSS subscribe to the principle that the amount of financial aid granted a student should be based upon financial need. The CSS assist colleges and universities and other agencies in determining the students' need for financial assistance. Entering students seeking financial assistance are required to submit a copy of the Parents' Confidential Statement (PCS) or Student's Financial Statement (SFS) to the College Scholarship Service, designating Northwestern Michigan College as one of the recipients. The PCS or SFS form may be obtained from a secondary school counselor or the College Financial Aids Office.

Application Procedures for Financial Aid. 1. Check the proper box on your admission application to indicate your interest in financial assistance. Northwestern Michigan College will then send you its application for financial aid.

2. Pick up a PCS or SFS and the Basic Grant Application form from your high school counselor or Office of Financial Aids.

3. Complete all three forms. The confidential statement and Basic

Grant application are sent to the appropriate address for processing. The NMC financial aid application is returned to your high school counselor or College Registrar to verify your grade point average.

4. All forms must be completed and on file before any action is taken. If you have any questions, call (616) 946-5650, extension 549. If an incomplete application is submitted, it will be returned with incomplete items marked, and financial aid will be delayed.

Generally, notification of financial assistance is made during the month of May prior to the academic year of attendance. You will be officially notified by mail as to the amount of your award. A student receiving financial assistance is awarded one-third of the amount at each term's registration. These funds are released only at that time.

Freshman applicants should file by February 1 for consideration for the next academic year. Sophomores and transfer students should apply by March 1 for consideration for the next academic year. If application, confidential statement, and Basic Grant application are filed after the established deadlines, students will be given consideration, although the chances of receiving an award are much more doubtful. Students applying for other academic terms must have application, confidential statement, and Basic Grant application on file ninety days prior to the final registration for that respective term.

All Northwestern Michigan College financial aids are listed and described at the end of this chapter.

College Regulations

These regulations may be revised if necessary by action of the College Board of Trustees.

Grading System. Complete records of student work are kept by each instructor, and a report of final grades is provided to every student following each term. A permanent record of all final grades is maintained for each credit student, and transcripts are available to him from the Admissions and Records office for a small fee.

The following are standard grades at Northwestern Michigan College:

4.0—Outstanding
3.5—Excellent
3.0—Good
2.5—Above Average
2.0—Average
1.5—Below Average
1.0—Deficient
0.0—Failed

Extracurricular Eligibility. All students are eligible to participate in extracurricular activities. Special admission students, those dually enrolled in another school, are excluded from participation.

All elected officers of campus organizations, participants in intercollegiate athletics, and others in responsible positions must maintain a current term 2.00 grade point average to remain eligible. One term of grace may be granted after review by appropriate college officials.

To help safeguard academic integrity, campus organizations, through their constitutions and by-laws; may specify additional requirements.

Dean's List. Full-time students who do unusually well in their studies are recognized on the Dean's List, which is published at the end of each term. The list includes the names of students, taking 12 hours of credit or more, who have achieved a current average of 3.5 or higher.

Financial Aids

Most of the scholarships, grants, and loan funds have been donated to Northwestern Michigan College by Grand Traverse area individuals, organizations, and business firms. Other financial aids are made available to qualifying students through special funds established by federal and Michigan legislation. Procedures for students to follow in applying for financial aid were described earlier, in the section "Financial Information."

Supplemental Educational Opportunity Grants. Northwestern Michigan College, in cooperation with the federal government, makes Supplemental Educational Opportunity Grants available to a limited number of undergraduate students with exceptional financial need who require these grants to attend college. To be eligible, the student must also show academic or creative promise.

Eligible students who are accepted for enrollment on a full-time basis, or who are currently enrolled in good standing, may receive a Supplemental Educational Opportunity Grant for each year of their higher education, although the maximum duration of a grant is four years. Grants range from $200 to $1000 a year, and can be no more than one-half of the total assistance given the student.

Basic Educational Opportunity Grant. Students should complete the BOEG·application, which is available from colleges and high schools. Mailing instructions are on the form. Once the applicant receives the "Student Eligibility Report," he submits it to the institution or institutions where he wishes to apply. The college will then compute the amount of the Basic Grant. The program provides grants of up to $1400 minus the expected family contribution. Full-time, three-quarter time, and half-time students are eligible to apply.

National Direct Student Loans. Northwestern Michigan College participates in the National Direct Student Loan Program. High

School graduates who have been accepted for enrollment, or students enrolled in full-time courses who need financial help for educational expenses, are eligible for student loans.

An undergraduate student may borrow up to $1250 each academic year to a total of $5000. The repayment period and the interest do not begin until nine months after the student ends his studies. The loans bear interest at a rate of three percent per year, and repayment of principal may be extended over a ten-year period, except that the institution may require the repayment of no less than $30 per month.

Michigan Higher Education Assistance Authority. Recognizing the fact that many families will not qualify for financial aid on the basis of need, yet are not in a position to assume the complete cost of a college education, a guaranteed loan program has been devised by the State of Michigan and the federal government in cooperation with local banking institutions and credit unions. The program is administered by the Michigan Higher Education Assistance Authority. Under the program, students may borrow up to $1500 a year from a bank or other financial institution. Applications are available through participating banks or credit unions.

Other Types of Financial Aid

College Work-Study Program. Students who need a job to help pay for college expenses are potentially eligible for employment by Northwestern Michigan College under the federally supported Work-Study Program. Eligibility for employment under this program requires that a student demonstrate financial need.

Students may work up to fifteen hours weekly while attending classes. During the summer or other vacation periods when they do not have classes, students may work full time (forty hours per week) under this program. In summer employment, under the Work-Study Program, an eligible student could earn approximately $900, if needed.

Regular Student Employment. Students who wish employment on campus, but feel they don't have a financial need should register with the Career Planning & Placement Center on campus.

The Academy is looking for students who want to make a career aboard the ships that ply the Great Lakes and connecting waterways. While the Academy does not stress military routine, it does operate in a disciplined environment consistent with the discipline found aboard ship.

The Academy is the only freshwater school of its kind in the United States.

The Academy program of study begins during the summer and continues aboard Great Lakes commercial ships.

The program combines technical training with studies in the liberal arts and sciences. Graduates of the program will qualify to write the Coast Guard license examination for either the first class pilot's license or the third assistant engineer's license, and will receive an Associate of Science degree and the preparation needed for employment with the Great Lakes maritime industry.

Through experience, following graduation, and with proven ability and performance, the graduate will have the opportunity to advance to master or chief engineer on the ships of the Great Lakes.

For additional information regarding costs, entrance requirements, the program of study and admission, address inquiries to:

Director of Admissions & Records
Northwestern Michigan College
1701 East Front Street
Traverse City, Michigan 49684.

APPENDIX

This appendix contains *formal* letters of application for nominations to the Service Academies.* For information about application, acceptance, and enrollment to the State Maritime Academies see the individual sections of Chapter 6 for the addresses.

1. Requesting a nomination to the U.S. Merchant Marine Academy.
2. A format letter requesting nomination to attend an Academy (Congressional).
3. Application for appointment as cadet, U.S. Coast Guard .
4. Request for presidential nomination.
5. Request for vice-presidential nomination.
6. Request for children of deceased or disabled veterans nomination.
7. Request for children of persons in a "missing status" nomination.
8. Request for children of Medal of Honor recipients nomination.

*The U.S. Coast Guard Academy differs from the other service academies in that appointments are made solely on the basis of an annual nationwide competition. A *formal* application form must be submitted by December 15. For the application form write to the Director of Admissions, U.S. Coast Guard Academy, New London, Conn. 06320.

Requesting a Nomination

(Sample Letter)

Date _____

The Honorable_____ or The Honorable_____
House of Representatives   United States Senate
Washington, D.C. 20515   Washington, D.C. 20510

Dear_____:

It is my desire to attend the United States Merchant Marine Academy. I respectfully request that I be considered one of your nominees for the class entering the Academy in the summer of 197___.

In the event that I am not selected for nomination to the Merchant Marine Academy, my alternate choices are numbered as to preference in the appropriate boxes below.

☐ United States Air Force Academy, Colorado Springs, Colorado

☐ United States Military Academy, West Point, New York

☐ United States Naval Academy, Annapolis, Maryland

The following personal data are provided for your information:

Full name _____
(Print as recorded on birth certificate)

Name of parents _____
Address: (Use ZIP code and phone number)
Permanent   **Temporary**
_____   _____
_____   _____
_____   _____
My date of birth: _____   Place of birth: _____
Social Security Number: _____
High School attended: _____
(Name and address)

My approximate standing is_____ in a class of _____

I have requested that a high school transcript of my work completed to date be forwarded to your office as soon as possible. I have also listed on the reverse side of this letter the results of any ACT or College Board test scores that I have taken.

I have been active in high school extracurricular activities as indicated on the reverse side.

I would greatly appreciate your consideration of my request for one of your nominations.

Sincerely yours,

(Signature)

1

Form Letters

Requesting a Congressional Nomination *(Sample letter)*

Honorable _____   Honorable _____
House of Representatives   or   United States Senate
Washington, D. C. 20515   Washington, D. C. 20510

Dear _____

It is my desire to attend the United States _____ Academy. I respectfully request that I be considered as one of your nominees for the class entering in the summer of 197___.

The following personal data are provided for your information:

Full name _____

(Print as recorded on birth certificate)

Name of parents _____

Address: (Use ZIP Code and phone number)

    Permanent   Temporary

    _____   _____

    _____   _____

    _____   _____

My date of birth: _____ Place of birth: _____

Social Security number: _____Home phone number:_____

High school attended: _____

(Name and address)

Date of high school graduation: _____ Sex: _____

My approximate standing is _____ in a class of _____.

I have/have not sent a Precandidate Questionnaire to the Academy.

I have requested my high school transcript of work completed to date be forwarded to your office as soon as possible. I have also listed on the reverse side the results of any ACT or College Board test scores that I have taken.

I have been active in high school extracurricular activities as indicated on the reverse side.

I should greatly appreciate your consideration of my request for one of your nominations.

Sincerely yours,

(Signature)

Notes: Prospective candidates should apply to their U. S. Representative and to both of their Senators.

             If you have not already filled one out, a Precandidate Questionnaire should be requested from the Director of Candidate Guidance, U. S. Academy, at the same time that your applications for Congressional nominations are submitted.

2

DEPARTMENT OF TRANSPORTATION
U.S. COAST GUARD
CG-4151 (Rev. 12-75)

APPLICATION FOR APPOINTMENT AS CADET, U. S. COAST GUARD
(Application must be postmarked not later than 15 December in order to be considered)
Form Approved OMB No. 04-R3019

INSTRUCTIONS: Complete this form and return to the Director of Admissions, U. S. Coast Guard Academy, New London, Conn. 06320. Read carefully before attempting to complete. Answer every question fully. This information will be kept confidential. Upon receipt of this application form CG-4151, the Academy will verify your eligibility and will then forward appropriate supplemental forms and instructions.

HOW DID YOU LEARN OF THE COAST GUARD ACADEMY? *(If more space is required, insert additional sheets as necessary)*

The United States Coast Guard reserves the right to investigate all statements made in this form. Any false statements or failure to disclose any material fact is sufficient cause for rejection of the application or dismissal after appointment.

I have read the scholastic and physical requirements for appointment as a Cadet in The U. S. Coast Guard and to the best of my knowledge and belief I meet all of these requirements. I will furnish proof of this statement upon request.

I CERTIFY on honor that the above statements are true and complete to the best of my knowledge and belief.

_____   _____
DATE   SIGNATURE

I VOLUNTARILY AUTHORIZE THE RELEASE OF THE INFORMATION CONTAINED THEREIN AND ANSWERS TO THE QUESTIONS CONCERNING MY APPLICATION TO THE FOLLOWING PEOPLE:

☐ YES ☐ NO MY PARENTS, GUARDIAN, OR RELATIVES
☐ YES ☐ NO GUIDANCE COUNSELOR OR SCHOOL OFFICIALS
☐ YES ☐ NO OTHER ADVOCATES IN MY BEHALF
☐ YES ☐ NO ALUMNI OR PARENTS INVOLVED IN APPLICANT FOLLOW-UP

_____   _____
DATE   SIGNATURE

SOCIAL SECURITY NUMBER

NAME *(Last, first, middle)* (PLEASE PRINT)

MAILING ADDRESS *(No. and Street)*

CITY   STATE   ZIP CODE

FOR ACADEMY USE ONLY

DISTRICT CODE | STATE CODE

TELEPHONE NO. *(Area Code)* | DATE OF BIRTH *(Month, day, year)*

PLACE OF BIRTH *(City, State)*

CITIZENSHIP
☐ U.S. NATIVE
☐ NATURALIZED
☐ FOREIGN

MARITAL STATUS
☐ MARRIED   ☐ WIDOWED
☐ SINGLE   ☐ DIVORCED

HAVE YOU BEEN DETAINED, ARRESTED, SUMMONED INTO COURT, INDICTED FOR CONVICTED FOR ANY VIOLATION OF CIVIL OR MILITARY *(Including Juvenile)* OFFENSES?   ☐ NO   ☐ YES

(Do not report minor traffic offenses)

(If "YES", insert a complete statement from a court official or probation officer giving nature of offense, date and disposition of case).

ARE YOU NOW OR WILL YOU BE A HIGH SCHOOL GRADUATE BY NEXT JUNE 30TH   ☐ NO   ☐ YES

(Give date of graduation _____)

PREVIOUS EDITIONS ARE OBSOLETE

Requesting a Presidential Nomination   *(Sample letter)*

(This application should be submitted after 1 June of the year preceding desired year of entry.)

To: Superintendent, U. S. _____ Academy, ATTN: Candidate Guidance Office

I request a Presidential nomination to the United States _____ Academy for the class which will enter in the summer of 197___ and submit the following data:

Name: _____
(Give full name as shown on birth certificate, or, if changed, attach copy of court order.)

Address: *(Use ZIP Code and phone number)*

Permanent	Temporary
_____	_____
_____	_____
_____	_____

Date of birth: _____   Social Security number _____
        (Spell out month)   *(Must be filled in)*

Name of high school: _____ Month/year of graduation: _____

Home telephone number: _____ Sex: _____

If member of military, check box ☐. List rank, serial number, component, branch of service and organizational address on reverse side of this form.

Information Concerning Parent's Military Service.

Name of parent: _____
        (Parent's rank, serial number, component, and branch of service)

I intend to request nominations from the following Members of Congress: _____

_____

_____.

The number of the Congressional district in which I plan to apply for a nomination is the _____ located in the state of _____.

Sincerely yours,
(Signature)

Note: In establishing your eligibility for a Presidential nomination, you should determine which of the following three service-connected categories applies to your parent, and forward the appropriate documents and information to the _____ Academy along with your letter of application for a nomination.

☐ Active duty officer: (Attach statement of service prepared by personnel officer specifying all periods of active duty.)

☐ Active duty enlisted: (Attach statement prepared by personnel officer specifying all periods of active duty and listing dates of enlistment and dates of expiration of enlistment.)

☐ Retired or deceased: (Furnish date and copy of retirement order or casualty report. If appropriate, include brief statement concerning the date, place and cause of death or the details of disability together with the Veterans Administration claim number. If eligible, applicant will be given a nomination in the Sons of Deceased or Disabled Veterans category.)

4

FORMAT

Request for Vice Presidential Nomination

Date.................................

The Vice President
United States Senate
Washington, D.C. 20501

Dear Mr. Vice President:

It is my desire to attend the   Academy and to serve in the United States . I
respectfully request that I be considered as one of your nominees for the class that enters the Academy
in June and submit the following data:

Name: (print as recorded on birth certificate) ....

Social Security number: ....

Permanent address: (street, city, county, state, zip code) ....

....

Temporary address: ....

Permanent phone number and area code: ....

Current phone number and area code: ....

Name of father: ....Name of mother: ....

Date and place of birth (spell out month): ....

....

Name and address of high school: ....

Date of graduation: ....Approximate grade average: ....

Extracurricular activities (Include athletic and non-athletic activities and work experience): ....

....

State your reasons for wanting to enter the   Academy: ....

....

I (have) (have not) received a prospective candidate questionnaire from the Air Force Academy.

I will greatly appreciate your consideration of my request for a nomination to the Air Force Academy.

Sincerely,

Signature ....

5

FORMAT

Request for Children of Deceased or Disabled Veterans Nomination

Date.................................

Director of Cadet Admissions
Academy

Dear Sir:

It is my desire to attend the   Academy and to serve in the United States. I request a nomination under the Children of Deceased or Disabled Veterans category for the class that enters the Academy in June and submit the following data:

Name: *(print as shown on birth certificate; if different from the name you use, attach a copy of court order, if applicable)* ....

Social Security number: ....

Permanent address: *(street, city, county, state, zip code)* ....
....

Temporary address: ....

Permanent phone number and area code: ....

Current phone number and area code: ....

Date and place of birth: *(spell out month)* ....
....

Date of high school graduation: ....

If member of military *(list your rank, social security number, regular or reserve component, branch of service, and organizational address including PSC and box no.)* ....
....

If previous candidate: *(list year and candidate number)* ....

Information on Parent

Name, rank, social security number, component and branch of service: ....
....

Date and place of death or date and place disability occurred: ....
....

Cause of death or disability: *(forwarding a copy of casualty report or copy of disability retirement order may expedite processing of your application)* ....
....

Veterans Administration XC claim number: ....

Address of VA office where case is filed: ....

Sincerely,

Signature................................

6

FORMAT

Request for Children of Persons in a Missing Status Nomination

Date...

Director of Cadet Admissions
Academy

Dear Sir:

It is my desire to attend the _____ Academy and to serve in the United States. I request a nomination under the Children of Persons in a Missing Status category for the class that enters the Academy in June and submit the following data:

Name: *(print as shown on birth certificate; if different from the name you use, attach a copy of court*

order, if applicable) ....

Social Security number: ....

Permanent address: *(street, city, county, state, zip code)* ....

....

Temporary address: ....

Permanent phone number and area code: ....

Current phone number and area code: ....

Date and place of birth: *(spell out month)* ....

....

Date of high school graduation: ....

If member of military *(list your rank, social security number, regular or reserve component, branch of*

service, and organizational address including PSC and box no.) ....

....

If previous candidate: *(list year and candidate number)* ....

Information on Parent

Name, rank, social security number, component and branch of service: ....

....

(Attach copy of DD Form 1300, Report of Casualty)

Sincerely,

Signature..................................

7

FORMAT

Request for Children of Medal of Honor Recipients Nomination

Date...

Director of Cadet Admissions
Academy

Dear Sir:

It is my desire to attend the _____ Academy and to serve in the United States. I request a nomination under the Children of Medal of Honor Recipients category for the class that enters the Academy in June and submit the following data:

Name: *(print as shown on birth certificate; if different from the name you use, attach a copy of court order, if applicable)* ....

Social Security number: ....

Permanent address: *(street, city, county, state, zip code)* ....
....

Temporary address: ....

Permanent phone number and area code: ....

Current phone number and area code: ....

Date and place of birth: *(spell out month)* ....
....

Date of high school graduation: ....

If member of military *(list your rank, social security number, regular or reserve component, branch of service, and organizational address including PSC and box no.)* ....
....

If previous candidate: *(list year and candidate number)* ....

Information on Parent

Name, rank, social security number, component and branch of service of parent to whom the Medal of Honor was awarded: ....
....

Sincerely,

Signature...

8

INDEX

INDEX

Aldrin, Edwin, 1
Anders, William, Jr., 38
Annapolis (*see* the United States Naval
    Academy)
The United States Air Force Academy, 67
    Airmanship, 83
    An Academy is Established (Early History),
      68
    Athletic program, 87
    Aviation Science, 85
    Graduation, 90
    Guide to, 69
    Mission and Objectives, 72
    The Academy Program, 70
    Questions and Answers, 92
    Research at Academy, 76
    Cadet Training, 77
    Underclass Years, 81
    Upperclass Years, 82
    Women in the Air Force, 75

Borman, Frank, 1
Burke, Admiral Arleigh A., 48

The California Maritime Academy, 210
    Academic Program, 228
    Academic Standards, 229
    Application, 231
    Career Opportunities in Today's Maritime
      Industries, 230
    Cost of Attendance, 216
    Curriculum, 219
    Extracurricular Activities, 226
    Facilities, 211
    Financial Aids, 217
    History, 210
    Location, 211
    Qualifications for Admission, 212
    Religion, 228
    Sea Training, 224
    Student Body Organization and Activities,
      234
Callaghan, Admiral Daniel, J., 48
Carter, James Earl, Jr., Midshipman, 39
Carter, Jimmy, President, 39
The United States Coast Guard Academy, 104
    Academic Departments, 120
    The Academy Philosophy, 127
    Academy Scholars Program, 124
    Cadet Allowances, 116
    How to Apply for Appointment, 127
    Individual Assistance and Counseling, 123
    Athletic Program, 118
    Coast Guard Duty, 124
    The Corps of Cadets, 115
    Courses of Instruction, 122
    What Classes are Like, 123
    A Day in the Life of a Cadet, 116
    Guide to, 105
    Early History, 104

Extracurricular Activities, 120
Fourth Class System, 115
Grades and Honors List, 123
Honors Placement, 123
Building Leadership Qualities, 113
Postgraduate Opportunities, 125
Professional and Military Training, 117
Questions and Answers, 128
Validation, 122
Visiting the Academy, 127
Women at the Academy, 116
Collins, Mike, 1

Dewey, Admiral George, 45

Eisenhower, Dwight D., 1

Farragut, Admiral David Glasgow, 43

Gilmore, Commander Howard W., 48
Grant, General Ulysses S., 1
Great Lakes Maritime Academy, 268, 275
    The Campus, 270
    College Regulations, 283
    The Curriculum, 271
    Entrance Requirements, 269
    Financial Aid, 280, 284, 285
    Learning and Working Flexibility, 278
Northwestern Michigan College, 274
    Student Activities, 279
    Student Health, 280

Harmon, Lt. Gen. Hubert R., 70

Jones, John Paul, 45, 46

Kincaid, Admiral Thomas C., 48
King, Fleet Admiral Ernest J., 45

Lee, General Robert E., 1
Lovell, James, 38

MacArthur, General Douglas, 1
McCampbell, Commander David, 48
The Maine Maritime Academy, 249
    Application and Admission, 251
    Athletic Programs, 264
    Extracurricular Activities, 258
    Financial Aid, 254
    History, 250
    Leave, 257
    Location, 250
    Placement, 265
    The Program, 249
    The Regiment of Midshipmen, 257
    Scholastic Preparation, 252
    Student Health, 256
    Transfers, 254
The Maritime College of State University
    of New York, 187

Academic Regulations, 205
Admissions, 198, 202
Athletic Program, 197
Buildings and Facilities, 188
Entrance Examinations, 199
Expenses, 202
Fort Schuyler, 188
Guide to, 188
Health Care, 196
History, 189
Location, 188
The Regimental System, 193
Religion, 196
Scholastic Requirement, 198
Student Life, 192
Student Organizations, 196
Transfers, 200
Undergraduate Academic Programs, 206
The United States Merchant Marine Academy,
    131
    Allowances and Expenses, 159
    Buildings and Facilities, 134
    Careers in the Maritime Industry, 152
    The Core Program of Education, 150
    Cultural Activities, 143
    Daily Schedule, 140
    Distinguished Academy Alumni, 155
    Extracurricular Clubs and Activities, 142
    General Requirements, 156
    Guide to, 131
    History, 132
    The Honor Concept, 138
    The Kings Point Sailing Squadron, 142
    Leave and Liberty, 140
    Location, 134
    Midshipman Life, 136
    Military Career Opportunities, 155
    Nominations, 156
    Plebe Orientation Program, 139
    Program of Study, 149
    Questions and Answers, 161
    Religion, 143
    Requesting Nomination, 156
    Student Loans, 160
    The U.S. Naval Reserve Midshipman
      Program, 145
    Varsity and Intramural Athletics, 141
    Walter P. Chrysler Estate, 132
    Women at the Academy, 133
The Massachusetts Maritime Academy, 170
    The Academic Program, 178
    Academic Standards, 180
    Advanced Placement Policy, 174
    Appointments, 176
    Cadet Life and Responsibilities, 181
    Cadet Programs and Activities, 185
    Career Counseling and Placement, 172
    College Level Examination Program, 175
    Financial Information, 177
    Guide to, 171

History, 170
Insurance, 185, 186
Intramural Athletic and Recreational
    Program, 185
Location, 172
Marine-related Career Opportunities, 173
Medical Care, 185
Minority Program, 175
Naval ROTC, 173
Scholastic Requirements, 173
Transfers, 176
U.S. Coast Guard Commission, 173
Varsity Intercollegiate Athletics, 183
Miles, Midshipman Alfred Hart
The United States Military Academy (West
    Point), 1
    The Academic Program, 13
    Admission of Women, 2
    Cadet Athletic and Extracurricular
        Activities, 21
    Early History, 7
    Guide to, 3
    The Future, 26
    Kosciuszko, Gen. Thaddeus, 6
    Location, 29
    The Military Training Program, 16
    Questions and Answers, 29
    Revolutionary War, 2, 4
    The Role of West Point, 10
    Thayer, Major Sylvanus, 9

The United States Naval Academy

(Annapolis), 37
An Academy is Established (Early History),
    40
The Academy Acquires Discipline, 41
Anchors Aweigh, 45
Challenges of Growth, 46
First Class Year, 55
Following Graduation, 61
Guide to, 38
June Week, 58
Questions and Answers, 62
Recent History, 56
Second Class Year, 54
Steps to Annapolis, 61
The Plebe Year, 49
Tecumseh, 44
Third Class Year, 53
Trident Scholar Program, 58
Nimitz, Admiral Chester, 48

O'Hare, Lt. Edward "Butch," 48

Pershing, Gen. John J., 1
Porter, Admiral David Dixon, 43

Schirra, Wally, 38
Shepard, Alan, 38
The State Maritime Academies, 165
Age Requirements, 169

The California Maritime Academy, 210
The Great Lakes Maritime Academy, 268
Health Requirements, 169
License Requirements, 168
The Maine Maritime Academy, 249
The Maritime College of the State University
    of New York, 187
The Maritime Industry, 166
The Massachusetts Maritime Academy, 170
The Texas Maritime Academy—Moody
    College of Marine Sciences and Maritime
    Resources—Texas A&M University, 232
U.S. Coast Guard Program, 168
U.S. Navy Officer Programs, 166

Texas Maritime Academy—Moody College of
    Marine Sciences and Maritime Resources,
    Texas A&M, 232
Academic Programs, 233
Admission, 238
Application, 240
Athletics, 244
Degree Programs, 235
Expenses, 241
History, 232
Student Activities, 243
Student Services (Financial Aids), 244
Tests Required, 240
Transfers, 240

West Point (see The United States Military
    Academy)